MW01517722

Rural Electrification

Hisham Zerriffi

Rural Electrification

Strategies for Distributed Generation

 Springer

Hisham Zerriffi
University of British Columbia
Vancouver, BC
Canada
hisham.zerriffi@ubc.ca

ISBN 978-90-481-9593-0   e-ISBN 978-90-481-9594-7
DOI 10.1007/978-90-481-9594-7
Springer Dordrecht Heidelberg London New York

Printed on acid-free paper

Springer is part of Springer Science+Business Media (www.springer.com)

For Conny and Aziza, who bring light to my world

Acknowledgments

The author conducted the underlying research for this book while he was a Post-Doctoral Scholar in the Program on Energy and Sustainable Development at Stanford University. He would like to thank the Program's director, David Victor, and the Program staff (past and present), particularly Rebecca Elias, Chi Zhang and Mark Thurber, for their support and help during this project. The author would also like to thank the World Resources Institute in Washington, DC, which provided him with a visiting fellowship during his research.

With a project of this size it is impossible to thank everybody who contributed in some way to the development and execution of the research. To those of you who took the time to meet with me, to provide your opinions and to help guide me on to the next person I should speak with, I sincerely thank you. As with all fieldwork-based research, this project would not have been possible without the assistance of a few individuals and organizations within each country who helped with introductions, logistics, translations, etc. In China, I am particularly indebted to Pan Jiahua and Peng Wuyuan for their time and energy. In Brazil, Adilson de Oliviera and his team at the Federal University of Rio de Janeiro, particularly Deborah Wanderley, were fabulous in setting up both this research and hosting a group of visiting students. In Cambodia, my work would not have been possible without the knowledge and assistance of the team from SME Cambodia.

Finally, the research and writing of this book would not have been possible without the support of numerous family and friends. It is hard to imagine that I would have had the energy and enthusiasm to make it through this project without them behind me. I am especially indebted to my partner, Conny, for her unwavering support and her ability to always keep me grounded.

Contents

Chapter 1
Rethinking Rural Electrification

Keywords Centralized versus distributed systems • Development • Distributed generation • Rural electrification

Introduction

Despite over a century of investment in electric power systems, there are roughly 1.6 billion people who lack access to electricity service, mainly in rural areas. While there are some open questions regarding the precise cause and effect relationships between rural electrification and human welfare, it is generally considered an important social, economic, and political priority to provide electricity to all. Unfortunately, the very complicated links between electricity and development are often obscured behind two idealized visions of rural electrification.

On the one hand is the image of the high-voltage transmission line, its tendrils reaching out into the countryside and bringing with it opportunities for jobs, communication, improved education, better health and a host of other welfare improvements (Fig. 1.1). Vaccines will be refrigerated, small industries can be made more productive, and children will be able to read their schoolbooks at night. This has been the traditional view of electrification: large power plants with long transmission and distribution lines. It is still the model favored worldwide by utilities and often, implicitly or explicitly, by regulators and policy-makers as well.

On the other hand is the "small is beautiful" image of a solar home system providing clean electricity for a remote farmer or household (Fig. 1.2). Many of the benefits are the same as with grid electricity: children reading their schoolbooks at night; farmers getting weather reports and market prices on the TV; household activities becoming easier and less labor intensive. This is how many in the NGO and international donor communities envision the future of rural electrification – environmentally sustainable and beneficial to the users. Many governments know that this imagery is also how they can get investment in their rural electricity projects, investments that would not occur without the linkage to sustainability and renewable energy projects in particular.

The problem with both of these images is that they are idealized visions of a much more complicated reality and fail to convey the complexities of solving the rural electrification problem. Rural electrification is a challenging task because it

H. Zerriffi, *Rural Electrification: Strategies for Distributed Generation,* DOI 10.1007/978-90-481-9594-7_1, © Springer Science+Business Media B.V. 2011

Fig. 1.1 Transmission towers at the Three Gorges Dam, China (photo: author)

involves delivery of a service to populations that are remote and dispersed and whose consumption is low. This means it is generally more expensive while at the same time the customer base is generally poorer and less able to pay the full cost of service. Combine these factors with utilities that are often poorly managed and have limited finances, and it is often not feasible to expect extension of the grid to unserved rural populations in the near future. Such conditions are also challenging for the development of new renewable energy technology markets.

In addition to the large number of people without any access to electricity, there are an untold number of people with access that is inadequate. Electricity is often limited to meeting the basic needs of households, and those basic needs tend to be in lighting and entertainment. Electricity for productive activities or for welfare enhancing community structures (e.g. schools or clinics) tends to lag behind basic household electrification or sometimes is completely neglected in rural electrification objectives, making integration of electrification into larger development goals difficult. Finally, electricity is often supplied by a wide variety of actors outside the traditional utility system, sometimes using technologies that are considered

Fig. 1.2 Cambodian home with solar panel on roof (photo: author)

undesirable from an environmental perspective (e.g. diesel engines). Without subsidies, the high cost of serving rural populations, either through the grid or through renewable and non-renewable off-grid options, results in high prices and consumption that is constrained by the ability and willingness to pay of the rural consumer.

The book is divided into three sections. Section I is an introduction to the general issue of electrification and rural development. The second section provides details on the research design and the results from the three country case studies. It also includes a chapter that compares and contrasts across the observations in all three countries to draw more general conclusions. The third and final section includes two thematic chapters that each cover an important issue in distributed rural electrification. The first is on universal electrification and equity. This chapter discusses the impact that both universal service goals and low rural electricity prices have had on rural electrification efforts. The second thematic chapter is on the financial needs for distributed electrification and discusses both the problem of access to capital and of cost-recovery.

Importance of Rural Electricity Supply

Access to electricity is considered a basic indicator of rural development, potentially contributing to income generation, improved educational and health outcomes, increased gender equality and a host of other social welfare improvements

(Goldemberg and Johansson 1995; World Bank 1996; WEC 1999; International Energy Agency 2004; Cabraal et al. 2005). These improvements come from both the direct benefits of electrification (e.g. higher productivity of agricultural producers due to use of electric motors and pumps) and from the indirect effects that come from access (e.g. improved knowledge of weather conditions and crop prices due to access to television and radio). However, the role that energy plays in development, either directly or in creating the enabling conditions for other development interventions to succeed, is still arguably underappreciated. For example, when the Millennium Development Goals were created as a way to highlight key development needs, energy indicators were not included explicitly. However, electrification will be necessary to meet many of the goals laid out in the MDGs, whether it is through refrigeration of vaccines (MDG4: Reduce Child Mortality) or lighting to improve evening study conditions (MDG2: Achieve Universal Primary Education). This led to a follow-on effort to delineate how energy contributes to the MDGs, but this was an ex-post effort rather than being integrated into the MDG development process (Modi et al. 2005).

Significant efforts have been made globally to provide electricity to both urban and rural populations. The global electrification rate went from 49% in 1970 to 73% in 2000, associated with 2.3 billion people gaining access during that time (International Energy Agency 2002). This represents a massive effort on the part of governments, international donor agencies, utilities and other actors. Much of this success can be attributed to the phenomenal success of China's rural electrification programs, particularly in the use of small hydro power. China went from having hundreds of millions without electricity in 1980 to less than 8 million today (IEA 2009). However, the gap between electricity needs and current levels of electrification remains large. Currently 1.6 billion people worldwide are unelectrified, primarily in rural areas. Even that figure, large as it is, leaves out the electricity necessary to contribute to broader patterns of rural development, since it only counts households and not income generating activities.

There are large regional disparities in electricity access, as shown in Table 1.1. South Asia and Sub-Saharan Africa have the lowest electrification rates while North Africa and the Middle East have reached greater than 90% total electrification. However, the rural/urban disparity within regions shows where the majority of those 1.6 billion people are located. Only in East Asia and North Africa are the rural electrification rates above 80%. The differences in Sub-Saharan Africa (8.4% rural versus 51.5% urban electrification), South Asia (32.5% rural vs. 69.4% urban) and Latin America (61.4% rural vs. 97.7% urban) are particularly striking. The rural electrification problem is challenging enough on technical grounds alone. The populations are often remote, sometimes in difficult terrain, and often widely dispersed. This makes the costs of grid extension high and is a major reason that distributed power generation alternatives are key to solving the problem.

For the majority of people without access to electricity from the grid, the most basic benefit of electrification, lighting, is provided by costly fuels like kerosene and candles, or indirectly via labor intensive traditional fuels such as wood or other biomass burned during cooking. These fuel sources can also have negative externalities,

Table 1.1 Urban, rural and total electrification rates by region, 2002 (From (International Energy Agency 2004)

Region	Population million	Urban population million	Population without electricity million	Population with electricity million	Electrification rate (%)	Urban electrification rate (%)	Rural electrification rate (%)
North Africa	143	74	9	134	93.6	98.8	87.9
Sub-Saharan Africa	688	242	526	162	23.6	51.5	8.4
Africa	**831**	**316**	**535**	**295**	**35.5**	**62.4**	**19.0**
China and East Asia	1,860	725	221	1,639	88.1	96.0	83.1
South Asia	1,396	390	798	598	42.8	69.4	32.5
Developing Asia	**3,255**	**1,115**	**1,019**	**2,236**	**68.7**	**86.7**	**59.3**
Latin America	428	327	46	382	89.2	97.7	61.4
Middle East	173	114	14	158	91.8	99.1	77.6
All developing countries	**4,687**	**1,872**	**1,615**	**3,072**	**65.5**	**85.3**	**52.4**
Transition economies and OECD	**1,492**	**1,085**	**7**	**1,484**	**99.5**	**100.0**	**98.2**
World	**6,179**	**2,956**	**1,623**	**4,556**	**73.7**	**90.7**	**58.2**

adversely affecting safety and indoor air quality. The use of televisions and radios or home appliances is either precluded or limited to low consumption appliances that use expensive disposable batteries. In rare cases, electricity will replace traditional biomass for cooking purposes. However, the high energy requirements for cooking and the cost of electricity usually result in either continued use of biomass or a transition to a more suitable modern energy source such as LPG.

Lack of electricity also affects the services provided by community buildings and local businesses. As with rural households, lighting for community buildings and for streets is seen as an important first use for electricity in rural communities. Another important use is refrigeration of medicines in local clinics. Access to electricity can also improve the productivity of local businesses – both in providing light to allow work to continue after dark, and in enabling the use of labor saving appliances and machinery.

Electrification can indeed enhance social welfare through augmented incomes, improved community health, and increased educational attainment. However, it should be noted that electrification, whether through the grid or by distributed means, cannot achieve these goals by itself and requires the presence of other enabling conditions (Barnes and Floor 1996; Martinot et al. 2002; Elias and Victor 2005). Without trained health and education professionals, there are limited gains that can be made by providing electricity to clinics and schools. Electrification of agricultural production can only increase incomes if the transportation system allows farmers to deliver their product to market in reasonable time.

The Contest to Electrify

Providing electricity in rural areas, particularly in developing economies, is challenging for three primary reasons. First, rural populations are usually dispersed and have low consumption, resulting in high capital costs spread over low returns. Second, the ability to pay of many rural populations is low, making centralized grid expansion unable to recover costs. Third, generation shortages, long rural feeder lines and poor maintenance often result in low quality power being delivered erratically to rural consumers.

Provision of electricity service worldwide has predominantly been through a large centralized system of power generation, transmission, and distribution. This system has developed over time due to the economies of scale provided by ever larger generating plants and the (perceived or real) natural monopoly characteristics of transmission and distribution. Accompanying the technical centralization of the power system has been an institutional centralization, with control over the systems resting with a small number of organizations (both governmental and private). Similarly, the regulation of these systems became centralized. Regulation has been limited to national or state level organizations or is implicit within the centralized utility itself. Even major efforts to restructure have mainly kept the technological, and much of the institutional, centralization of the system intact (Haugland et al. 1998; MacKerron and Pearson 2000; Victor and Heller 2006).

In many cases, this system has functioned relatively well. However, the economics of grid extension rely on spreading high costs over a maximized density of customers in a given region and a certain level of consumption. Ideally, those customers would all have the ability to demand and pay for electricity at the levels necessary to recoup costs. In reality, only a few or sometimes none of the customers in rural areas will be willing or able to pay the full costs of grid extension. However, universal electrification goals mean that grid extension will usually target all rural households, regardless of their demand level, willingness to pay or ability to pay. Even if it made economic or technical sense, it would be politically difficult for a centralized entity, often a state owned enterprise, to allow for differential electricity access within a small geographic area. As noted by Foley, this creates a "conflict of objectives" for utilities between financial performance and universal access and means that rural electrification is both a low priority and possibly a losing proposition (Foley 1992a).

Furthermore, centralized utilities in many countries simply do not have the managerial and financial resources to meet all rural electricity needs. Even in those areas where the grid does reach, electricity is often sporadic and of low quality, making it difficult to use for productive purposes or for vital tasks like vaccine refrigeration. For example, in India, voltage drops across rural feeders extend well beyond the limits considered acceptable by electrical engineers. Poor quality power can damage electrical equipment and result in more frequent loss of service (Tongia 2007). Even if these barriers could be overcome, it is not technologically and financially feasible or even desirable to use the grid to reach all rural customers.

Despite these challenges, with a few notable exceptions, universal access programs have generally been implemented through the centralized utility system.[1] Of course, centralized utilities do have an important role to play in rural electrification. For many rural customers, the grid is the lowest cost option available and will thus be the dominant mode of electrification. However, the high costs of extending grids to rural areas, the lower ability of rural customers to pay for electricity service, and the financial instability of many utilities in these regions mean that some rural areas will not gain access to the grid in the foreseeable future. There is nothing, however, in the theoretical goal of universal electricity access that requires it to be met through centralized utilities. However, a combination of regulations, historical path dependence and deep-seated norms have led to the predominant use of the centralized utility system for rural electrification. Utility responsibility has, at times, even been extended to cases where distributed generation technologies have been selected. Brazil provides the perfect example, with regulations dictating exclusive service territories for its utilities (thereby effectively eliminating competition) and allowing those utilities to meet their universal service obligations through a combination of grid extension and distributed generation sources.

---

[1] China has supported local utilities and renewables markets to meet its impressive service goals. Other countries, such as the Philippines, have relied on cooperatives to meet needs in some areas (Barnes 2007).

Given that utilities are often not interested in rural electrification due to the poor returns and technical difficulties, there are a number of possible options for removing rural electrification responsibility from the main utility.[2] (Foley 1992a, b) Traditionally, these have included establishing an autonomous division within the utility, creating a separate rural electrification agency, and devolving more responsibility to local organizations such as cooperatives and local communities (Foley 1992a). As this study shows, private and semi-private options have the potential to play a large role. Each option, including maintaining responsibility within the main utility, has implications for the use of distributed power generation. Not only may the approach chosen affect technological priorities and biases, but it may also affect access to resources and the various other institutional support mechanisms that are important for successful rural electrification.

Over the past few decades a number of small-scale, distributed power generation (DG) technologies that are suitable for rural areas have been developed or improved.[3] Distributed power generation is attractive for rural electrification for a number of reasons. For example, the low population densities and low consumption of rural customers are well matched to the scalability and autonomous operation possibilities of distributed power. Grid extension is expensive in rural areas and generally means trying to provide electricity that is available (in theory, at least) 100% of the time and at levels that may be much higher than typical rural consumption levels. In addition, rural customers will have an even greater imbalance than urban ones between their minimum and maximum (peak) loads. Many rural customers do not have refrigerators or other appliances requiring constant power and will use their electricity in the evening hours only for lighting, entertainment and, in some cases, cooking. While rural consumption is relatively low, its additive effect right at the time of peak power demand on the system can force the utility to run more expensive generating units more often or even to invest in new peaking generation. This can significantly raise the cost of supplying rural customers (Howells et al. 2006). Distributed power is able to provide power at levels and at times that are well-matched to rural customers. Finally, the possible set of organizational models is much larger with distributed power, including the possibility of decentralized local organizations (either private or public). This can alleviate some of the high transaction costs inherent in a centralized organization like a utility administering a customer base that is large and geographically diffuse (Hansen and Bower 2003; Chaurey et al. 2004; Banerjee 2006).

---

[2] Foley notes other institutional problems with utilities that are rural electrifiers, including the low prestige garnered among engineers for working on low-voltage systems, administration problems due to a diffuse customer base and their centralized nature which favors a small number of large projects (Foley 1992b).

[3] The definition of distributed generation is complicated and often context dependent. For the purposes of this book, electric power generation is considered to be "distributed" when it is produced locally and primarily consumed locally. See Appendix A for a more detailed discussion.

There are a number of ways that distributed power can provide rural electricity service. In addition to the various technologies that can be used (both fossil fuel based and renewable), there are several possible modes of installation and operation. Electricity generation technologies can be directly installed in homes, shops, factories and community buildings. Alternatively, service can be provided to end-users through a small grid system. For example, a DG owner could provide electricity to immediate neighbors, to a village mini-grid, or even to a multi-village local grid. A third option is to use the distributed power generation source for local battery charging. These batteries are often car batteries that can be used in households to power lights and small appliances.

Installed systems range widely in size and usage depending on technology choice, installation mode, and the institutional model chosen (see below). Some provide only enough power for basic needs, while others allow for higher consumption appliances like refrigerators, blenders, and sewing machines.

While DG technologies may be the best (or only) option in many circumstances, it must be recognized that there are also disadvantages to their use. Many of these technologies are more expensive than grid-generated electricity on a per kW basis and would not be competitive if the grid was eventually extended (or if existing grids were strengthened to provide reliable power) (ESMAP 2000; ESMAP 2005). When combustion engines are used, there are limited pollution controls (if any), contributing to both local and global environmental problems. Depending on the institutional model that resulted in the DG installation, there may also be little or no support for operations and maintenance, leading to shortened technology lifetimes (Nieuwenhout et al. 2001; Martinot et al. 2002).

Despite their inherent advantages in some contexts, the diffusion of these technologies and the supply of electricity in rural areas generally remains far below the technological and economic potential due to various institutional factors that existing studies of rural electrification tend to ignore. A systematic assessment of those institutional factors is one focus of this study.

The scope for distributed power generation on the one hand is a function of fundamental technical factors related to topography, population density, and the like that make centralized systems ill-suited to serving some rural populations. On the other hand, it is also a function of managerial and institutional factors related to the operation of the utility and its ability to expand, willingness to pay, investment incentive, and subsidies that change the relative economics of centralized versus distributed systems. The zone where centralized systems and distributed systems could both meet demand is therefore not fixed. In any system there will be a contestable area where both distributed and centralized solutions are competitive with one another for service provision. This can be seen in Fig. 1.3.

This contestable area is not fixed and can move in either direction, expanding or shrinking the space for distributed or centralized solutions. Technological changes that reduce costs will increase the area where DG is competitive. Alternatively, changes in the utility system that make it able to expand further into rural areas – for example, the provision of subsidies to the utility – will cause the contestable

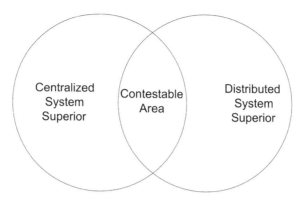

Fig. 1.3 The scope for centralized and distributed systems

area to impinge further into the domain of distributed generation. Conversely, limitations on grid extension into rural areas due to poor finances, mismanagement or simple neglect of these areas will effectively increase the space in which distributed systems are competitive.

This Study: Distributed Rural Electrification

While both centralized and distributed technologies are needed to tackle the rural electrification problem, this study focuses only on the options related to distributed electrification. Given the wide diversity of distributed electrification options that are available, a better understanding of how distributed electrification has been implemented in the past is necessary in order to guide future policy and investment decisions.

Prior Experience with Distributed Generation

The use of distributed power sources for rural electrification is not new. In a limited way, distributed power generation has been used for decades. In the 1970s, a large effort was made by international donors to provide off-grid technologies for rural electrification as well as to expand grids and solve other rural energy problems; that work has continued in one way or another to the present. As for the success of distributed power generation in providing rural electricity, the historical record is mixed at best (Barnett 1990; Martinot 2001; Martinot et al. 2002) Larger scale successes include Kenya's photovoltaic market and China's small hydropower systems (Acker and Kammen 1996; Duke et al. 2002; Pan et al. 2006).

There are more numerous examples of individually successful installations across a wide range of technologies and organizational models.[4] The use of diesel generators by Brazilian utilities, particularly in the Amazon, demonstrates the success of both that technology and a centralized mode of implementation. However, while diesel generators are widely used, their operation in remote areas by centralized utilities is not a model that has been widely replicated, and even in Brazil, their use by the utilities to power remote mini-grids remains a measure of last resort. Other localized success stories can be found worldwide. For example, a South African solar home system dissemination project had only limited success in installations and little success in its larger institutional goals. However, those systems that were installed did work and provide their customers with electricity as designed (Green et al. 2001). This is typical of many success stories, in which a relatively small number of units are installed successfully, but once the project is over, there is no further dissemination.

There are a few examples of broader success in establishing viable and self-sustaining distributed generation models that expanded beyond their initial, limited scope. One example of success is the use of small hydro stations in China, which now total more than 42,000 units with generating capacity reaching 28 GW (units less than 50 MW) (Tong 2004). Another often-cited success story is the photovoltaic market in Kenya, which originally started as a donor program and spawned a viable private industry (Acker and Kammen 1996). Similarly, there are PV cash markets in a number of other countries, some selling tens of thousands of units per year. The literature on distributed rural electrification tends to focus on the success of renewables, particularly wind and photovoltaics. However, diesel generators have also played a large role in many places and should be considered a successful technology from an electrification point of view (World Bank 1996; Enterprise Development Cambodia 2001; ESMAP 2005).

There have also been numerous failures in using distributed power generation in rural areas. These failures occurred at the installation or project level as well as in translating individual small successes into larger ventures. In some cases, particularly in the 1970s, failure was simply due to problems with the technology itself. Arguably, technical failure in the early stages is not an inherent problem and provides opportunity for learning. However, if a distinction is not made between experiments and programs for wide-scale diffusion, such technical failures can cause serious problems for further diffusion efforts (Barnett 1990). In other cases, the technology itself was successful, but the institutional mechanisms were not put in place to sustain the system over a longer period of time. Early solar home system programs in which it was assumed that the technology was essentially maintenance free are one example.

---

[4] Here we use the rather limited definition of "success" to mean that the installation meets the expectations of the parties involved in terms of cost and service, and the technology remained operational for a reasonable amount of time. A more nuanced and precise definition of success is used to assess projects in the actual analysis.

There are also cases in which projects move beyond the technology demonstration phase and are successful in providing electricity within a given area. However, there has been a general failure to capitalize on that success and translate it into a replicable model for rural electrification. The earlier focus on technology demonstration projects meant that even when the technology worked, there was no effort to create and demonstrate a viable model for further diffusion or the necessary structure for maintenance, financing and continued operation (Martinot et al. 2002). A recent example is the set of solar home system programs funded by the Global Environment Facility (GEF). While projects in individual countries have been successful in providing solar home systems to thousands of households, they have not become self-sustaining and replicable, despite some resources being put towards institutional strengthening. Each additional set of installations requires international donor funding and coordination, and the scale of installations remains small compared to the need.

While it is necessary to apply technologies that are appropriate for the context, it is the institutional factors which determine whether a particular implementation is successful and whether it is replicable and sustainable over the long term. In 1992, Foley noted the tendency to focus on the technology (often emphasizing the suitability of renewables) while ignoring the important questions regarding institutional arrangements for installing, operating and maintaining the system (Foley 1992a).

While there has been an increasing emphasis on institutional factors, there continues to exist a technology focus in rural electrification projects, with insufficient attention to institutions and particular biases in favor of renewable technologies. For example, government programs for provision of renewable energy systems continue to create obstacles to the development of sustainable technology markets for long-term diffusion. As noted recently, there is still a tendency among donors to provide large capital cost subsidies or even to donate equipment, despite the history of such programs being unsustainable and harming the creation of viable markets (Martinot et al. 2002).

Zimbabwe is a perfect example of this phenomenon. Through an international donor program dozens of local enterprises for photovoltaic systems were founded in the mid-1990s. However, due to their dependence on the donor funds and lack of provision for a transition to a more market-oriented approach or sustainable support structure, the enterprises all failed once the donor program ended. This not only meant that new systems were no longer being sold, but also that service and maintenance of existing installations disappeared (Martinot et al. 2002).

These problems have been exacerbated by issues of technological lock-in, in which choices made now restrict the range of choices possible in the future. Such path dependencies are a familiar problem in technology diffusion and can often occur for technical reasons (economies of scale in production, long capital lifetimes, etc.). However, lock-in can also occur for institutional reasons as lessons fail to be learned and donors and governments find it difficult to adapt. As Barnett notes, "the process of technology diffusion often requires such a strong commitment to a particular device that the leadership is reluctant to admit that there are problems" (Barnett 1990).

The State of Research on Distributed Electrification

Unfortunately, the existing literature on successes and failures in distributed electrification is only partially helpful as a guide for general policy-making. We can divide the literature into three broad categories.

First are the technology specific analyses. These are the numerous reports, articles and books that focus on the opportunities (and sometimes the challenges) of using a particular technology to meet rural electricity needs. Almost invariably the technologies examined are renewable energy. In some cases, it is the broad category of renewables (Allderdice and Rogers 2000), while in others it is a specific renewable energy source or technology, such as biomass or photovoltaic systems (van Campen et al. 2000; Li et al. 2001).

Second are the micro-level project reports. These report on a particular activity, usually within a few years after implementation. This category includes numerous village level projects ("technology X was installed in village Y and worked/ failed"), as well as reports on broader programs covering a larger region (Green et al. 2001; Santos and Zilles 2001; Stroup 2005). These are useful for understanding some of the micro-level factors that contributed to that particular success or failure (though generally it is only successes that are reported). However, a broader understanding is only possible by aggregating such individual project experience. As with the first category, Technology X is almost always a renewable energy technology. Non renewable sources are generally included only as a baseline against which to contrast the renewable technology.

Third are the business success stories. This literature is a start towards filling a major gap in the literature, namely the need to understand why business models for distributed electrification succeed or fail. However, far too often this category overlaps with the first and what is reported is relevant only for a specific technology, such as photovoltaics. More importantly, the studies in this category tend to only focus on successes, often reporting on "best practices" for a technology or type of project (ESMAP 2001).

What all these studies have in common is their ad-hoc approach in studying a limited set of previous projects to determine what did and did not work for providing electricity to rural uses (Hurst 1990; Erickson and Chapman 1995; Allderdice and Rogers 2000; Martinot et al. 2002; Etcheverry 2003; Fishbein 2003). While these studies provide some useful information, they can suffer from a case selection bias since their scope is limited in geography (one country, region or even village), technology (only PV or only wind or only renewables), or end-use (household electrification, productive uses). Often they also suffer a bias in selecting "success" cases while ignoring the failures. As will be discussed in the next section, it is difficult to avoid some of these pitfalls and no study could examine every distributed rural electrification effort undertaken. However, this study attempts to avoid systematic biases by not selecting or rejecting cases a priori on the basis of technology, end-use or outcome.

This study also fills a gap in previous work by addressing many of the institutional issues that are known to impact outcomes in rural electrification through a carefully constructed case-based analysis. Many of the individual case studies discussed above do include discussion of institutional issues such as regulations, electrification policies, access to financing, etc. However, for the same reasons as above, it is difficult to generalize because of their scope. On the other hand, there is prior literature that covers many of the institutional issues addressed in this research. The findings in this literature have been generalized from the secondary literature and from the authors' admittedly extensive experience (Barnett 1990; Foley 1992a; Barnes and Floor 1996; Radulovic 2005; Reiche et al. 2006). One exception is recent work by Barnes that looks at rural electrification programs across a number of countries (Barnes 2007). However, while providing valuable information and comparative analysis, the analysis is not focused on the role of distributed electrification options as in this work.

Research Methods Used in This Study

There are many countries in which some form of distributed electrification has been attempted. From this universe of countries, we have chosen to focus on three: Brazil, Cambodia and China. These three countries have very different institutional environments (particularly in their regulatory and policy regimes) and different business models for distributed rural electrification – in fact multiple business models within each country. In all, there are roughly 20 different models across the three countries. We exploit the variation between the models in each country and the variation between the institutional contexts of the three countries in order to examine factors important for success and failure.

Based upon a review of the literature, discussed above, four independent variables were chosen to capture the important elements of the **business models used for distributed rural electrification**: Organizational Form, Technology Choice, Target Customers and Financial Structure. The Organizational Form variable looks at whether the primary organization responsible is centralized or decentralized and whether it is governmental or non-governmental. The Technology Choice variable categorizes business models according to whether they use renewable versus non-renewable energy technologies and whether the system is a mini-grid or individual installations. The Target Customers variable is used to examine how models that electrify households perform differently than those that electrify productive activities or community structures. The Financial Structure variable provides information on how capital is obtained and how operational costs are covered.

While the variables described above cover the important aspects of the business models themselves, they do not include some key information for understanding outcomes. Two categories of data were added to the study in order to capture institutional factors and physical context dependent factors.

Data relevant to the institutional context were added as control variables. The level of capital and operating subsidies were given scores on a low-medium-high scale. In addition to subsidies, there was a need to categorize the policy and regulatory regimes more generally in order to capture the impact of the institutional context on the distributed electrification models. The policy and regulatory regimes were characterized as favorable, neutral or unfavorable.

Data relevant to the physical context were also included as control variables. We are particularly interested in the remoteness and the density of the population. Remoteness bears directly on the potential viability of grid extension and on the potential difficulties related to project management and operations and maintenance. The density of the population is relevant for the relative viability of the grid, micro-grids and individual installations.

Each business model was assessed based on three main dependent variables: Changes in Electricity Service, Sustainability and Replicability. Changes in Electricity Service primarily measures the increase in electricity access as a result of the business model. Secondary measurements are of the sufficiency and quality of the electricity supplied. Sustainability is primarily a measure of the ability of the model to cover its costs and provide functioning systems over a long period of time. Replicability is a measure of whether the particular characteristics of the business model can be used to provide electricity services to new customers. Together these three dependent variables measure the short and long-term impact of a business model on the electricity supply situation. The dependent variables used to assess the performance of the business model for distributed rural electrification were scored on a High-Medium-Low scale according to a set of pre-specified criteria.

Data for the study were collected through a combination of secondary sources, site visits and interviews. In particular, officials within relevant ministries and regulatory authorities were interviewed as well as donors, academics and representatives of non-profit organizations. This provided valuable information about the history of electrification efforts and the institutional context for rural electrification. Interviews and site visits were used where possible to collect information about specific distributed electrification efforts and to supplement information from secondary sources. More details on the research methods can be found in Chapter 2.

Summary of Results

Brazil's distributed rural electrification experience has been dominated by highly centralized efforts. This includes government ministry programs for electrifying community buildings and utility run programs for household electrification. The historically slow pace of rural electrification, however, has led to a number of models outside the utility and government systems, including solar home system leasing schemes, cooperatives, private diesel micro-grids and renewable energy based systems for agricultural producers.

The Cambodian situation differs remarkably from the Brazilian one. Official electrification rates remain low (15%) and the state utility only serves the largest population centers. Unofficially, electricity is widespread as a result of rural electricity entrepreneurs that run diesel based micro-grids, battery charging stations or a combination of the two. Cambodia also has a small solar home system market that serves slightly wealthier consumers and has had some donor projects.

China has had stunning success in rural electrification since the early eighties. Small hydropower has played a major role and currently provides power for over 300 million rural Chinese (more if one includes systems with only partial hydropower supply). The government was heavily involved, but often in a supporting role through low interest loans and guarantees and technology programs. Technology development and support has also been key to China's success in creating rural markets for solar home systems and small wind and wind/pv hybrid systems. Recently the Chinese government instituted the Township Electrification Program, which was a much more top–down centralized effort.

This study of distributed rural electrification in Brazil, Cambodia and China shows that a wide diversity of distributed rural electrification models exist and can be successful in providing electricity to remote rural populations. However, while their individual successes and failures are not always due to the same set of factors, there are some common themes and clear trends that emerge. An important conclusion of this study is that, in the absence of strong central support for rural electrification, alternative electrification models (e.g. private diesel operators, cooperatives, NGOs providing alternative energy) emerge to meet the needs of different consumers. Lacking financial support from the central government, successful models have had to meet requirements for financial sustainability in other ways. These independent efforts tend to serve a customer base exhibiting the following characteristics:

- Users include productive activities or other high energy consumers (e.g. coops and NGO projects in Brazil).
- Relatively wealthier (e.g. PV customers in Cambodia, wind and hybrid customers in China).[5]
- Willing to pay very high prices for very low consumption (e.g. unlicensed diesel genset customers in Cambodia).

Another point that emerges from the study is that all distributed technologies can be used successfully. There have been and continue to be situations in which technologies are inadequately piloted before wider distribution, are manufactured poorly or have other technological shortcomings. In the great majority of recent cases, however, technology implementations have failed more for institutional reasons than technical ones. In the absence of outside support to introduce renewable technologies, local technology choice will tend towards diesel generation (e.g. diesel mini-grids in Cambodia and Brazil, battery chargers in Cambodia). Renewable energy technologies have generally relied on regional, national or international

---

[5] These are customers that are at the top of the "base of the pyramid." The base of the pyramid, a term covering the vast majority of the population that is usually ignored by commercial enterprises due to assumptions of their low buying power, has become a powerful organizing idea for creating new opportunities to make money while solving societal problems and meeting environmental goals. See, for example, Hart (2005).

institutional support for introduction, product improvement and market improvements (e.g. wind, PV and small hydro in China, PV in Brazil). However, this can become a problem when renewables introduction goes hand in hand with a technology dump approach.

Main Conclusions of the Study

In many countries, the question is not whether distributed generation has a role to play. Rather it is a question of *how* it will play a role. This study set out to look at the historical experience with DG for rural electricity supply in order to answer a couple of fundamental questions:

- How can DG systems be installed and run in way that is financially sustainable and replicable and in a way that meets the needs of rural populations?
- What is the role of the institutional context in determining choices in technologies and business models?

Wider use of distributed electrification in a manner that meets local needs requires a new vision, one that moves beyond a focus on basic household electrification and on particular technologies. Electrification should be based on the diversity of local needs and decision-making processes, including the need for electricity to improve productive activities. At the same time, there remain good reasons for regulator oversight, and new regulatory mechanisms have to take into account the particular nature of distributed systems (Reiche et al. 2006).

The financing of the business model emerges as one of the key issues to be resolved in the use of distributed electrification in rural areas. Providing technology for free has proven to be unsustainable and difficult to replicate. If subsidies are to be used, they should be primarily for helping overcome the high capital costs of technologies. However, the effects of subsidies can be both positive and negative and care has to be taken in developing subsidy programs. They can force a more centralized solution and undercut options that might be better suited to contribute to overall development in rural communities. New and innovative financing schemes need to be developed and have the opportunity to be tested.

The results of this study point to five major conclusions that bear on efforts to provide distributed rural electrification and reach a goal of sustainable universal service and improved rural development. We have cast these conclusions as major lessons learned:

Lesson 1: Observations Are Not Analysis

There is a clear need to move beyond analysis based primarily on anecdotal evidence or single-N studies (e.g. one place, one technology or one business model). More formalized and rigorously developed research designs are necessary in order to develop better policies and inform actors in the distributed rural electrification field. The good news is that the data needed to conduct such analyses are available or can be developed through careful study design (e.g. surveys, semi-structured interviews, etc.). This study was based on work in three countries and examined roughly 20 business models. Similar work could be done in more countries and could address a broader set of questions regarding rural energy patterns.

Lesson 2: Free Lunches Are Not Sustainable (And They Kill the Restaurant Industry)

There are a number of factors that determine success and failure in distributed rural electrification. One clear lesson is that all technologies can work and various organizations can be successful (or can fail) depending on local conditions, institutional context, and a host of other factors. However, one factor which is clearly important across all cases is the need to have some level of cost-recovery and financial sustainability. If subsidies are to be used to ease the burden of high costs, they should focus primarily on keeping down first costs, and they should be carefully considered and designed before implementation. Subsidy programs that result in free electricity are usually unsustainable over the long-term and can prevent the implementation of options that are ultimately more sustainable and replicable.

Lesson 3: Electrons Do Not Equal Development

Electrification on its own does not guarantee development, either from an income generation perspective or in improving social welfare outcomes. The focus on pushing electrons is due in part to the tyranny of indicators in international development activities. In the case of rural electrification, that indicator is generally the number of households electrified. Rather than focusing on household electrification to the exclusion of other rural needs, it is important for maximum impact to improve electrification in the context of larger development patterns. This also means that the emphasis needs to change from *whether* a given technology can provide electricity to *how* a given technology will supply electricity. It is necessary to stop pilot testing the technologies themselves and begin rigorously pilot testing and evaluating institutional models for implementing those technologies. This has to be coupled with an approach that looks at overall rural energy needs, particularly for income generating activities, not just household needs. It is also vital to overcome the bias towards promotion of renewables by outside actors. The contribution of rural communities to climate change is minimal and they should not be forced to bear the burden of mitigation at the expense of development. Renewables and more conventional generation sources should compete equally to meet rural development needs, taking into account capital costs, maintenance costs, fuel costs and variability, local institutional factors, local environmental factors and a whole host of other parameters.

Lesson 4: Think Globally, Act Locally

The stock phrase of the environmental movement applies equally well when thinking about what it takes to realize large-scale electrification that contributes to rural development. Centralized organizations, particularly government energy ministries and large utilities, focus primarily on pushing electrons and not on rural development. They have a difficult time understanding local conditions, which are key determinants of whether an electrification program is successful and furthers the development objectives of the community. The involvement of local actors provides greater versatility in meeting rural electrification goals, draws upon a wider range of actors, and results in a greater diversity of activities being undertaken. Rural electrification is an important country-wide and global objective, but it is often best achieved through local means.

Lesson 5: Unbias the Social Contract

The push for rural electrification flows from the social contract that governs the relationship between the state and its people. Due to the economies of scale usually associated with electricity, fulfillment of that social contract has tended to be biased towards centralized organizations, typically ministries or utilities, as the agents of the state. This study shows the need to unbias the social contract and open the system up to a wide variety of actors that can provide service on a local level. In other words, the provision of services should be performance based, not size based. This type of institutional change may be difficult in some places and has to be done in a way that minimizes the increase in transaction costs that can occur from more decentralized action.

Appendix A: Defining Distributed Generation

There is, unfortunately, no agreed upon definition of distributed generation (DG). Small generators fired on diesel and natural gas, as well as small renewables (hydro, solar, wind, etc.) are presumably the most familiar DG technologies. The two fuels least likely to be associated with distributed power are coal and uranium. However, in China a large number of coal-fired power plants were built in the 10 MW range which served primarily local distribution networks. While nuclear power reactors for civilian purposes have tended to be large centralized installations for a variety of reasons, the use of nuclear reactors onboard ships shows the degree to which they can be made compact and indeed, plans for future generations of nuclear reactors have included smaller reactors in the tens of MWs, which would be suited to power supply through distribution systems.

Formal definitions have tended to be highly context dependent and focus on one or more particular characteristic of either the technology or its use. (For a detailed discussion of how to define distributed generation see Pepermans, Driesen et al. 2005 and Ackermann et al. 2001). Often, distributed generation is defined according to the generation technology and fuel (at least implicitly). This leads to the assumption among some that DG implies the use of renewable technologies or, conversely, the use of small natural gas or diesel engines. In fact, nearly every source of electricity generation can be (and has been) made small enough to be considered "distributed." Ownership is sometimes used to define DG, usually to mean that the unit must be owned by the end-user. However, this precludes a number of possible institutional arrangements, such as utility co-ownership or energy service company ownership. A third characteristic often used is the operational mode – whether power generation is dispatchable or can be scheduled.

All three of these characteristics are limiting and result in an overly narrow definition of distributed generation. Ackermann et al. provide a useful discussion of these and other characteristics often used to define distributed power generation (Ackermann et al. 2001). Their conclusion is that distributed generation can only be defined as generation that is located within the distribution network or "on the customer side of the meter" (which accounts for both off-grid and on-grid applications). They then place various qualifiers to account for other characteristics (e.g., micro vs. small vs. medium vs. large distributed generation). While this definition is very flexible, and allows for technologies of different sizes, operation modes, and purposes to be included, it is too expansive for the purposes of this study.

This paper adopts a narrower and more precise definition. Generation is considered "distributed" if the power is generated and consumed locally (this would be close to what Ackermann et al. define as "embedded generation," wherein the power output is used only within the local distribution network). This definition allows for significant flexibility in technologies and operation modes as well as institutional arrangements. Technologies can range from solar home systems to diesel engines. These can be operated completely off-grid or as part of a local mini-grid. The "local distribution network" includes under this definition completely off-grid technologies such as solar home systems where no network actually exists. However, as noted, the usefulness of this definition is in limiting power generation and consumption to the local level. Ownership can include individuals, groups of individuals, communities, private commercial actors, and national governments. Given that this study focuses on the use of distributed generation to meet the power needs of rural populations, this more limited definition of embedded generation is appropriate.

References

Acker RH, Kammen DM (1996) The quiet (energy) revolution: Analysing the dissemination of photovoltaic power systems in Kenya. Energy Policy 24(1):81–111

Ackermann T, Andersson G et al (2001) Distributed generation: a definition. Electric Power Sys Res 57:195–204

Allderdice A, Rogers JH (2000) Renewable energy for microenterprise. National Renewable Energy Laboratory, Golden, pp 1–70

Banerjee R (2006) Comparison of options for distributed generation in India. Energy Policy 34:101–111

Barnes DF (ed) (2007) The challenge of rural electrification: strategies for developing countries. Resources for the Future, Washington DC

Barnes DF, Floor WM (1996) Rural energy in developing countries: a challenge for economic development. Annu Rev Energy Env 21:497–530

Barnett A (1990) The diffusion of energy technology in the rural areas of developing countries: a synthesis of recent experience. World Dev 18(4):539–553

Cabraal RA, Barnes DF et al (2005) Productive uses of energy for rural development. Ann Rev Env Resour 30:117–144

Chaurey A, Ranganathan M et al (2004) Electricity access for geographically disadvantaged rural communities: technology and policy insights. Energy Policy 32(15):1693

Duke RD, Jacobson A et al (2002) Photovoltaic module quality in the Kenyan solar home systems market. Energy Policy 30:477–499

Elias RJ, Victor DG (2005) Energy transitions in developing countries: a review of concepts and literature, PESD Working Papers. Program on Energy and Sustainable Development, Stanford University, Stanford, p 38

Enterprise Development Cambodia (2001) Survey of 45 Cambodian rural electricity enterprises. Enterprise Development Cambodia, Phnom Penh, p 108

Erickson JD, Chapman D (1995) Photovoltaic technology: markets, economics and rural development. World Dev 23(7):1129–1141

ESMAP (2000) Mini-grid design manual. Energy Sector Management Assistance Programme, Washington DC

ESMAP (2001) Best practice manual: promoting decentralized electrification investment, Energy Sector Management Assistance Programme.

ESMAP (2005) Brazil background study for a national rural electrification strategy, Aiming for Universal Access. Energy Sector Management Assistance Program (ESMAP), Washington DC, p 176

Etcheverry J (2003) Renewable energy for productive uses: strategies to enhance environmental protection and the quality of rural life. Department of Geography and Institute for Environmental Studies. University of Toronto, Toronto, 49.

Fishbein RE (2003) Survey of productive uses of electricity in rural areas. Africa Energy Unit, Washington DC, pp 1–50, World Bank

Foley G (1992) Rural electrification in the developing world. Energy Policy

Foley G (1992b) Rural electrification: the institutional dimension. Utilities Policy 2(4):283–289

Goldemberg J, Johansson TB (eds) (1995) Energy as an instrument for socio-economic development. UNDP/BDP Energy and Environment Group, New York

Green JM, Wilson M et al (2001) Maphephethe rural electrification (photovoltaic) programme: the constraints on the adoption of solar home systems. Development Southern Africa 18(1):19–30

Hansen CJ, Bower J (2003) An economic evaluation of small-scale distributed electricity generation technologies. Oxford Institute for Energy Studies, Oxford, p 66

Hart SL (2005) Capitalism at the crossroads: the unlimited business opportunities in solving the world's most difficult problems. Wharton School Publishing, Upper Saddle River

Haugland T, Bergesen HO et al (1998) Energy structures and environmental futures. Oxford University Press, Oxford

Howells M, Victor DG et al (2006) Beyond free electricity: the costs of electric cooking in poor households and a market-friendly alternative. Energy Policy 34(17):3351–3358

Hurst C (1990) Establishing new markets for mature energy equipment in developing countries: experience with windmills, hydro-powered mills and solar water heaters. World Dev 18(4):605–615

IEA (2009) The Electricity Access Database. International Energy Agency. http:www.iea.org/weo/database_electricity/electricity_access_database.htm

International Energy Agency (2004) Energy and development. In: World energy outlook 2004. International Energy Agency, Paris, p 38

International Energy Agency (2002) Energy and poverty. In: World energy outlook 2002. International Energy Agency, Paris, p 47

Li J, Xing Z et al (2001) Biomass energy in China and its potential. Energy Sustain Dev V(4):66–80

MacKerron G, Pearson P (eds) (2000) The international energy experience: markets, regulation and the environment. Imperial College Press, London

Martinot E (2001) Renewable energy investment by the World Bank. Energy Policy 29:689–699

Martinot E, Chaurey A et al (2002) Renewable energy markets in developing countries. Annu Rev Energy Env 27:309–348

Modi V, McDade S, et al (2005) Energy services for the millennium development goals. Energy Sector Management Assistance Programme, United Nations Development Programme, UN Millennium Project, and World Bank, New York

Nieuwenhout FDJ, van Dijk A et al (2001) Experience with solar home systems in developing countries: a review. Prog Photovoltaics Res Appl 9:455–474

Pan J, Peng W, et al (2006). Rural electrification in China 1950–2005. Program on Energy and Sustainable Development Working Paper #60. Research Centre for Sustainable Development, Chinese Academy of Social Sciences and Program on Energy and Sustainable Development, Stanford University, Beijing, China and Stanford, CA

Pepermans, Driesen et al. (2005) Distributed generation: definition, benefits and issues. Energy Policy 33:787–798

Radulovic V (2005) Are new institutional economics enough? promoting photovoltaics in India's agricultural sector. Energy Policy 33:1883–1899

Reiche K, Tenenbaum B et al (2006) Promoting electrification: regulatory principles and a model law. Energy and Mining Sector Board, Washington DC, p 49, World Bank

Santos RRd, Zilles R (2001) Photovoltaic residential electrification: a case study on solar battery charging stations in Brazil. Prog Photovoltaics Res Appl 9:445–453

Stroup KK (2005) DOE/NREL inner Mongolia PV/wind hybrid systems pilot project: a post-installation assessment. National Renewable Energy Laboratory, Golden, p 106

The Electricity Access Database. International Energy Agency. http://www.ica.DrWeo/database.electricity-access-database

Tong J (2004) Small hydro power: China's practice. China WaterPower Press, Beijing

Tongia R (2007) The political economy of Indian power sector reforms. In: Victor DG, Heller TC (eds) The political economy of power sector reform: the experiences of five major developing countries. Cambridge University Press, Cambridge, pp 109–174

van Campen B, Guidi D et al (2000) Solar photovoltaics for sustainable agriculture and rural development. Food and Agriculture Organization, Rome, pp 1–76, United Nations

Victor D, Heller T (eds) (2006) The political economy of power sector reform: the experiences of five major developing countries. Cambridge University Press, Cambridge

WEC (1999) The challenge of rural energy poverty in developing countries. World Energy Council and Food and Agriculture Organization of the United Nations, London, p 199

World Bank (1996) Rural energy and development: improving energy supplies for 2 billion people. World Bank, Industry and Energy Department, Washington DC

Chapter 2
Research Design

Keywords Distributed rural electrification models • Research methods • Study design

Introduction

The history of distributed rural electrification efforts makes it possible to develop a structured study in order to understand the factors important for both success and failure. In order to do so, it is necessary to be specific regarding the characteristics of the distributed electrification model believed to be important in determining outcomes, the outcomes to be measured and the other factors that can be influential. This chapter outlines the methods used in this study, the hypotheses developed to explain outcomes, and how the particular case studies were chosen.

The distributed rural electrification efforts examined in this study vary according to the types of organizations involved, the technologies chosen, their financial structure and the customers targeted for electrification. These variables are hypothesized to have a direct impact on changes in electricity service and the financial performance of rural electrification efforts. The choices made along these four dimensions create a distinct organizational and business model for using distributed generation for rural electrification. We have coined these models Distributed Rural Electrification Models (DREMs), and each DREM is one observation in our study.

The institutional factors affecting the deployment of different DG systems range from regulatory aspects, such as rules governing licensing of generation facilities, to much more normative and cognitive constraints that affect access to financing and technical support mechanisms. These factors are complex and varied, operating at many different levels to create constraints and incentives for the choice of technology and organizational form for providing rural electricity. The institutional factors also affect the performance of DG systems once they have been deployed.

The research relies on a sample of case studies that is large enough to span wide regional variation yet small enough to allow in-depth analysis – a so-called "medium-N" approach. This approach involves selecting a sample of potential case study countries for which there is variation along dimensions thought to be key to

H. Zerriffi, *Rural Electrification: Strategies for Distributed Generation*, DOI 10.1007/978-90-481-9594-7_2, © Springer Science+Business Media B.V. 2011

the outcomes. Cases are chosen along two dimensions. First, they are selected for between-country variation in terms of economic, regulatory and policy environment. Second, within each country particular examples of distributed rural electrification programs are selected on the basis of within-country variation in terms of technologies and institutional arrangements. The end result is a set of approximately 20 cases consisting of individual distributed rural electrification programs or projects within a small set of countries. This medium-N approach allows us to systematically choose case study countries and individual cases in a way that makes use of limited resources while meeting requirements for variation and reducing case study selection biases.[1] The countries selected are China, Brazil and Cambodia and more detail on the case selection process is provided at the end of this chapter.

This chapter is divided into two sections. The next section covers the variables used in the study. The variables used to measure immediate outcomes of distributed electrification efforts are described and the criteria used for scoring the variables are provided. This is followed by detailed descriptions of the four parameters used to describe the Distributed Rural Electrification Models. Hypotheses on how each DREM parameter impacts electrification outcomes are also provided. The final section describes how the cases and observations were selected. The overall case selection methodology is outlined followed by more detailed information on each of the countries selected for the study.

Study in Context

Here we briefly describe how distributed electrification efforts relate to both policies and improved social welfare outcomes. Policymakers often adopt policies that encourage the supply of modern energy services such as electricity. Through a long causal chain, such policies are thought ultimately to affect household welfare (see Fig. 2.1). This research focuses on one important link in this causal chain: the organizational and business model through which distributed rural electrification takes place. The ultimate goal of our research in this area is to eventually understand how to develop policies that maximize social welfare.

Changes in energy supply in rural areas, particularly the move to modern energy services such as electricity, can have wide ranging impacts on social welfare. The move away from kerosene or candles for lighting can improve health due to reduced indoor air pollution. Adequate lighting can allow children to study in the evenings, raising education levels. Electric motors can be used to improve the productivity of small industrial or agricultural producers. Electricity can also help to facilitate the emergence of small household enterprises – often run by women. This final causal link between energy and welfare (marked as Causal Link III in Fig. 2.1) is not directly addressed in this research.[2]

---

[1] For an excellent review of methods for social science research and the power of case study approaches see King et al. (1994).

[2] The relationship between energy and welfare is the subject of other ongoing projects. For a review of the welfare impacts of electrification see Ramani and Heijndermans (2003), Pachauri and Spreng (2004), Cabraal et al. (2005), and Elias and Victor (2005).

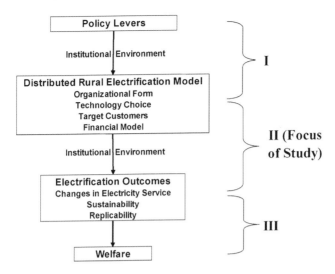

Fig. 2.1 The policy to welfare causation chain

Causal Link I is between the policy environment and the business models chosen. Policy incentives that favor one type of organization or make certain types of financing easier to use will have an impact on the business model chosen. Additionally, various institutional factors come into play that can shape and alter those choices. By creating a set of rules, both formal and informal, institutions create an environment in which distributed rural electrification must operate. That environment, along with factors such as technical resource availability (e.g. amount of sunlight in the case of PVs), shapes and constrains the choices that are made and the ability of actors in the system to meet their objectives. It should be noted that this institutional environment can act to help or hinder the adoption of particular organizational models or technologies.

This study focuses in the middle – on the link between characteristics of the Distributed Rural Electrification Model (DREM) and the performance of different DG systems (Causal Link II). The parameters of the DREM include the nature of the organization undertaking distributed rural electrification, the technologies chosen, the target customer base, and the financial structure with regard to capital and operational costs (all of these factors are discussed further below). The impact of the DREM chosen on the immediate outcomes of distributed rural electrification are mediated by a set of institutional factors that must be accounted for in explaining the outcomes.

Variables for the Study of Distributed Electrification Business Models

Based upon a review of the literature, discussed in Chapter 1, four independent variables were chosen to capture the important elements of the **D**istributed **R**ural **E**lectrification **M**odel (**DREM**): Organizational Form, Technology Choice, Target Customers and

Financial Structure. The Organizational Form variable looks at whether the primary organization responsible is centralized or decentralized and whether it is governmental or non-governmental. The Technology Choice variable categorizes DREMs according to whether they use renewable versus non-renewable energy technologies and whether the system is a mini-grid or individual installations. The Target Customers variable is used to examine how models that electrify households perform differently than those that electrify productive activities or community structures. The Financial Structure variable provides information on how capital is obtained and how operational costs are covered.

Each DREM was assessed based on three main dependent variables: Changes in Electricity Service, Sustainability and Replicability. Changes in Electricity Service primarily measures the increase in electricity access as a result of the DREM. Secondary measurements are for the sufficiency and quality of the electricity supplied. Sustainability is primarily a measure of the ability of the DREM to cover its costs and provide functioning systems over a long period of time. Replicability is a measure of whether the particular characteristics of the DREM can be used to provide electricity services to new customers. Together these three dependent variables measure the short and long-term impact of a DREM on the electricity supply situation.

Data for the study were collected through a combination of secondary sources, site visits and interviews. In particular, officials within relevant ministries and regulatory authorities were interviewed as well as donors, academics and representatives of non-profit organizations. This provided valuable information about the history of electrification efforts and the institutional context for rural electrification. Interviews and site visits were used where possible to collect information about specific distributed electrification efforts and to supplement information from secondary sources.

The dependent variables used to assess the performance of the Distributed Rural Electrification Models were scored on a High-Medium-Low scale according to a set of pre-specified criteria (discussed below). For example, the diffusion of solar home systems by the centralized utility, COELBA, is given high scores for access, sustainability and replicability. This program will diffuse roughly 30,000 solar home systems and will be the primary way in which COELBA meets the electricity needs of its customers it cannot reach by the grid, accounting for its high score on the Access variable. Due to the ability of COELBA to cross-subsidize its service and its obligations under the regulatory system, this model is given a high score on the sustainability parameter since the utility can reasonably be expected to continue its service. Replication of the solar home system program beyond the initial phase with expectations of full service to all households is evidence that replication has been widespread and so this variable is also scored as high. It should be noted that, in this example, the utility is able to take advantage of favorable policy and regulatory regimes and subsidies for capital (through government grants) and operating expenses (through cross-subsidies). This is reflected in the control variables and it is therefore possible to see that this model's outcomes rely upon those favorable regimes and subsidies and that, absent those conditions, the model would not be successful (since there would not be the financial mechanisms to make it viable without significantly changing the model).

Outcomes of Distributed Rural Electrification

The purpose of this study is to better understand the influence of different distributed rural electrification models on three principal outcomes: changes in electricity service, sustainability and replicability. Each of these three outcomes is described below. Outcomes are measured on a scale of Low-Medium-High. Specific criteria for assigning these scores are provided and in each of the case study chapters, justification will be provided for assigning a particular value to the outcome for each DREM (See Boxes 1–3 and Table 2.1).

Changes in Electricity Services

The stated goal of both grid-connected and distributed rural electrification efforts is to improve the level of electricity service available to rural populations. There are four key dimensions by which the improvements in electricity service can be measured.

The most common indicator of improved electricity service is an increase in access to electricity by rural populations; this will be the main indicator used to measure changes in electricity services in this study. Access refers to the presence of a source of electricity, whether from the grid or a distributed power source. Definitions of electricity access differ according to context; for example, until recently, villages in India were considered electrified if they contained one or more connection to the electricity grid (Tongia 2003). One problem is that statistics on electricity access are usually at the household or population level and do not account for access by commercial enterprise or public buildings in rural areas. Official statistics also usually count only official connections to licensed or approved distributors. Illegal connections, unlicensed distributors and off-grid sources may not be counted.

However, access may not always be enough to meet the energy needs of rural populations. Even if a connection or power source is available, the energy provided may be insufficient (either due to capacity constraints or to frequent outages).[3] As a secondary measure of change in electricity service, we consider the sufficiency of the electricity supplied by the DREM. As with availability, there are differing definitions of sufficient electricity supply. The structure of the local economy, including the types of rural enterprise and income level of the population, determine the amount of electricity that is sufficient. In some cases, minimal target levels of electricity consumption by rural households have been set (Tong 2004). As with availability, commercial and productive activities are usually not considered when assessing sufficiency of electricity supply (with the possible exception of small agricultural

---

[3] It should be noted that in some cases, supply may actually exceed community demand, at least in the short term. This could happen if a generating unit is sized too large due to a miscalculation of the populations' willingness or ability to pay or if the unit is sized to meet some productive application during certain hours of the day and then is oversized for the load of the local population when used for community and residential purposes.

Box 1 Measuring Changes in Electricity Service

Assigning scores to changes in electricity access is difficult as the changes must be assessed against a baseline. We have chosen not to have an absolute baseline in numeric terms (i.e. number of people served) as this would unfairly skew results when making inter-country comparisons due to large population differences. All baselines are therefore intra-country baselines and a DREM that has a high score in Cambodia and one that has a high score in China should be considered together even if one serves a few million and the other one hundreds of millions. Even within a country, an absolute baseline is not justified given that the resource potential might be different. Put simply, given the hydrological resources of China and its solar insolation, more people can be served at a lower cost using small hydro than solar systems. However, there are some areas where solar systems make more technical and financial sense than hydro systems. To consider the Chinese cash market for PV systems to have had a low impact because it did not reach the hundreds of millions of people supplied by small hydro does not provide an adequate assessment of how well it performed in meeting rural electrification needs in the north and northwest of the country where small hydro is not an option. Therefore, scores are given on the basis of whether it has been extended beyond the pilot phase and whether it is the dominant form of distributed service delivery in that area. Scores for sufficiency and quality are somewhat easier to derive. However, data on sufficiency and quality are harder to obtain. In general, therefore, access will be the primary measure used for electricity service (despite the problems detailed above). Sufficiency and quality will be discussed when appropriate as supplemental measures of electricity service.

	High	Medium	Low
Access	It is the dominant mode of service delivery in that area and has extended beyond the pilot phase	It has extended beyond the pilot phase but is not the dominant mode of service delivery	Occurs in a handful of communities
Sufficiency	Enough power is available to meet general demands and there is little or no exit from the system	Enough power is available to meet general demands but the system is run at full capacity and/ or some portion of customers exit the system	Enough power is available only to meet basic demands (e.g. lighting and one low-consumption appliance in the case of households) even if customers require more
Quality	Outages approach those of the main grid utilities and power fluctuations and line voltage drops are not a major issue	Outages are higher than the main grid and power quality is lower but long outages (> 1–2 days) and damage to equipment are rare	Frequent longer outages, high voltage drops over mini-grid lines and damage to equipment is common

Box 2 Measuring Sustainability

The key metric for sustainability is continued electricity service to the target customers. In order for a DREM to score high on sustainability, that service must be maintained without major changes to the basic features of the model. If service is maintained, but the model required major changes (e.g. changes in ownership, addition of a new tariff to cover costs), then the original model is not considered to be sustainable (though the new model may be). Failure times of 5 years and 2 years are considered to have low and very low scores on sustainability.[4]

High: Continued performance up to the expected lifetime of the technology is demonstrated or reasonably expected without major changes to the basic model.

Medium: Continued performance up to half the expected lifetime of the technology is demonstrated or reasonably expected without major changes to the basic model.

Low: Failure to continue to deliver electricity beyond 5 years or major changes required to the model in order to continue electricity beyond 5 years.

Very Low: Failure to delivery electricity beyond 2 years after installation.

Box 3 Measuring Replicability

Replicability is measured primarily in terms of new installations within the same institutional context (in this case, within the same country).

High: Only marginal changes required to either the financial structure or institutional arrangements in order to replicate and evidence of actual replication.

Medium: Some changes required but relatively adaptable.

Low: Significant changes would have to be made to the business model in order for it to be replicated. This can be the result of failure of the original model, reliance on specific financial resources that may not be widely available or reliance on institutional arrangements that are unique and difficult to change.

pumpsets in some countries). In this study, both availability and sufficiency of electricity access will be considered for household as well as non-household needs.

A third dimension of electricity service is supply reliability and quality. Electricity supply that is either erratic in terms of reliability (the amount of time

---

[4]Five years is chosen because this is slightly longer than the replacement time for a number of components on some of the key technologies (e.g. batteries for solar home systems) and it is reasonably expected that most technologies would have to undergo some sort of repair or parts replacement. The Very Low score is included with a 2 year cutoff to account for the fact that historically, some projects have had very fast failure rates (failing even before the expected lifetime of the shortest-lived replaceable component).

Table 2.1 Summary of Criteria used to score dependent variables

	High	Medium	Low
Electricity access	It is the dominant mode of service delivery in that area and has extended beyond the pilot phase	It has extended beyond the pilot phase but is not the dominant mode of service delivery	Occurs in a handful of communities
Sufficiency	Enough power is available to meet general demands and there is little or no exit from the system	Enough power is available to meet general demands but the system is run at full capacity and/or some portion of customers exit the system	Enough power is available only to meet basic demands (e.g. lighting and one low-consumption appliance in the case of households) even if customers require more.
Quality	Outages approach those of the main grid utilities and power fluctuations and line voltage drops are not a major issue	Outages are higher than the main grid and power quality is lower but long outages (> 1–2 days) and damage to equipment are rare	Frequent longer outages, high voltage drops over mini-grid lines and damage to equipment is common
Sustainability[a]	Continued performance up to the expected lifetime of the technology is demonstrated or reasonably expected without major changes to the basic model	Continued performance up to half the expected lifetime of the technology is demonstrated or reasonably expected without major changes to the basic model	Failure to continue to deliver electricity beyond 5 years or major changes required to the model in order to continue electricity beyond 5 years
Replicability	Only marginal changes required to either the financial structure or institutional arrangements in order to replicate and evidence of actual replication	Some changes required but relatively adaptable	Significant changes would have to be made to the business model in order for it to be replicated. This can be the result of failure of the original model, reliance on specific financial resources that may not be widely available or reliance on institutional arrangements that are unique and difficult to change

[a]The Sustainability metric also includes a score of Very Low to account for those cases in which the DREM fails almost immediately (i.e. within 2 years after installation). These are generally cases where technology failure occurs quickly and the DREM is not able to provide for service

electricity is unavailable) or quality (voltages and frequencies that deviate from accepted operating parameters) can have a serious impact on electricity service. Unreliable supply can negate some of the potential advantages of electricity access (e.g. the use of refrigeration for food and medical supplies). Poor quality can cause damage to equipment such as irrigation pump motors. However, it is necessary to distinguish between lower reliability (or quality) that is planned versus random. For example, take a household receiving electricity only 4 h/day. That may be because the household only demands electricity in the evening. The household may not be willing or able to pay for 24/7 service. On the other hand, it could be because the grid fails regularly and randomly, despite a desire, willingness to pay and ability to pay for 24/7 service. Clearly, those two households will view their 4 h/day service quite differently.

The fourth dimension of electricity service is affordability. Both distributed technologies and grid extension into rural areas can require large capital investments and ongoing operation and maintenance costs. Depending on the amount of electricity demanded and the socio-economic profile of the users, payment of up-front capital costs can place a heavy burden on rural households and small commercial enterprises. Monthly operations and maintenance costs, on the other hand, are often not a completely new and additional cost in the household budget. The new electricity service often displaces other energy sources (e.g. if candles are being used for illumination or dry cell batteries are used to power radios) that are both expensive and of lower quality. In some cases, electrification can result in a reduction in monthly household energy expenditures or the provision of higher quality services for the same cost.

These changes in electricity service have to be assessed on both a short and long-term basis in order to determine the impact of the Distributed Rural Electrification Model. Some models may result in significant short-term changes, particularly in terms of access, but may not continue to provide access over the long-term (e.g. some large solar home system donation programs) (Nieuwenhout et al. 2001). Others may only have the resources to provide access to a small group over the short term, but may establish a model that can eventually result in long-term access for a larger population. However, for sake of clarity and to simplify measurement, this variable will only be used to measure *short-term* changes in electricity service. Long term effects of a DREM on electricity service are captured in both the sustainability and replicability measures. A sustainable DREM will mean that whatever short-term changes result in electricity service will continue for those customers. A replicable DREM implies that over the long term, those short-term gains will be achieved for a wider population.

Sustainability

In order for any distributed rural electrification program to be considered successful it must be able to install and maintain systems over the long-term. In the context of rural electrification, financial performance can be difficult to sustain given that some rural populations are unable to pay market prices for electricity. In addition, there are often political imperatives for electrification arising from the expected relationships between electrification and development. As a result, subsidies of some form are often involved. However, this does not mean that financial performance of the different models cannot be evaluated and compared. The subsidy (both amount and nature of the subsidy) becomes a factor in evaluating the model.

While financial sustainability is our primary focus in this part of the research, to the extent possible we will also take into account non-financial factors that can affect program sustainability. These non-financial or institutional factors include the presence of trained maintenance personnel, the availability of fuel and ease of fuel transport, regulations that either hinder or improve operational efficiency or cost-recovery tariff levels, etc.

Replicability

In order for any DREM to contribute to serving long-term rural electrification needs and improving welfare it must be replicable. Often, it has proved difficult to expand projects, even when they are considered successful (Acker and Kammen 1996). Replication of models beyond their initial base or to entirely new contexts (e.g. a different country) can fail for a number of financial and institutional reasons, including unreliable provision of government support, lack of commercial financing, and reliance on fine-tuned policies or regulations that are difficult to replicate in new contexts.

Replicability is even more difficult to measure than sustainability for two reasons. First, it is possible for a model to be *replicable* without actually having been *replicated*. Therefore, for some cases judgments will be made regarding the *potential replicability* of the model. Second, unless an explicit reference to a previous DREM is made in project documents, it is not always possible to trace the replication of a model. All models will require some degree of adjustment in order to fit into a new context and it may not always be possible to determine if a model was replicated. As a result, replicability will be a secondary benchmark used in evaluating outcomes. When possible, conclusions will be drawn regarding replicability of the DREM, but changes in electricity service and sustainability of the DREM are the main dependent variables to be measured.

The long-term sustainability of a particular distributed rural electrification effort and its replicability are related concepts. A precise definition of sustainability and replicability is difficult. For the purposes of this study, they are defined as follows:

> A Distributed Rural Electrification Model is considered to be both sustainable and replicable if it can continue to meet the needs of its initial users and meet the demand of a new set of similar users without significantly changing the financial or institutional parameters.

For example, if the government subsidizes an initial installation of 5,000 solar home systems but cannot afford to subsidize the remaining 300,000 households at the same level, necessitating a change in the tariff to a level considered unacceptably high, the project may be sustainable (the initial 5,000 installed systems continue to function), but is not replicable (there is no way to scale up from 5,000 to 300,000).

Independent Variables

There are a number of possible business models that can be adopted to provide DG-based electricity in rural areas (Foley 1992; Martinot et al. 2002). Based on a review of the literature, we identified four fundamental characteristics of the business models that determine success in distributed rural electrification efforts: the organizational nature of the enterprise; the target customer base; technology choice and the financial structure (including financing of capital, subsidies, and tariff structure). Each of these characteristics of the DREM will be described more fully below. After each one, a set of hypotheses will be developed regarding the impact of that DREM characteristic on the outcomes of electricity service, sustainability and replicability. Unfortunately, there is little literature to guide the development of some of the hypotheses but, where possible, references will be made to prior studies. For sake of clarity, hypotheses will be developed based on each characteristic, assuming all aspects of the other characteristics are held constant. For example, in developing hypotheses for the impact of the financial structure, the type of organization, the target customers and the technology choice are assumed to be the same. However, it is not always possible to have complete separation of the variables. For example, financial structures that rely on cross-subsidies are more likely to be implemented by centralized utilities rather than decentralized organizations.

Organizational Form

One important characteristic of a business model is the nature of the organization that is the locus of technology diffusion. There are numerous characteristics that can be used to define an organization. In this study we have divided organizations along two dimensions. The first is whether the organization is centralized or decentralized; the key determinant being whether decisions are made at the point where the technology is being implemented. The second is whether the organization is governmental (local, state or national) or non-governmental (NGOs, cooperatives and commercial enterprises).

The organizations involved in distributed rural electrification can range from highly centralized (e.g. national governments, large companies) to decentralized (e.g. individual owners, local entrepreneurs, local governments). We ask, do distributed systems perform better when the implementing organization is itself decentralized? Or can economies of scale in organization (reduction of transaction costs and greater resources) result in higher performance by centralized organizations?

A distinction can also be made at both the local and higher levels between governmental and non-governmental organizations. The question of the merits of privatization of infrastructure is longstanding and ranges across infrastructure services (water, sewage, transportation, etc.). In the electricity sector, until recently many countries have had wholly government-owned utilities. Market reforms have led to a reduction in the role of central government (usually through the development of independent power producers and occasionally through the privatization of distribution systems). And the arguments in favor of privatization of electricity supply at the national level are also a question for rural electricity supply. These include the efficiency of operations and of the potential of competition to reduce costs and improve service to rural populations.

The same actor may be involved in more than one business model, playing a slightly different role depending on the nature of the model. For example, the national government can implement a completely centralized program in which its ministry or state utility provides the funds, determines technology choices, and implements the program. At the same time, that government can also be supporting a decentralized, non-governmental effort through information dissemination, support for financing organizations and establishment of favorable tariffs and subsidies. In the latter case, the actual purchase, installation and maintenance of the technology are done at a local, non-governmental level. In both cases, the central government is involved, however only the first case can really be considered a case of centralized governmental diffusion based on the criterion stated above regarding whether decisions are being made at point of implementation. By this standard, the second case would be best characterized as an example of decentralized, non-governmental activity. Table 2.2 shows some of the organizations that would have the combination of centralization and governmental nature discussed above. This is not meant to be an exhaustive list, but rather demonstrates what is meant when referring to organizational form.

There are, of course, variations on these combinations of different types of organizational models. For example, technology sales may be through a large centralized entity (e.g. Honda selling gasoline or diesel generators through its dealer network) but the technology purchaser may be a local entrepreneur who uses that technology for a local electricity service business (e.g. a small mini-grid serving shops in a marketplace).

Centralized/Governmental Organizations

At the national and state or provincial level, governments have a long history of involvement in distributed rural electrification. In some cases, this has been institutionalized into government ministries or other administrative entities. India

Table 2.2 Examples of distributed rural electrification organizations

	Centralized	Decentralized
Governmental	National or State Governments	Community Councils
		County Governments
Non-Governmental	International Donor Organizations	Entrepreneurs
	Technology Suppliers	Cooperatives

is a good example of this as it has established a Ministry of New Energy Sources (MNES) – which is involved in both grid-connected and off-grid renewable energy – as well as the Rural Electrification Corporation and the Indian Renewable Energy Development Agency.

The involvement of governments at these levels can sometimes be relatively passive, with the intention of providing technical support funds for distributed rural electrification without being the planner or implementing agent. In other cases, state/provincial and national governments have been directly involved in planning and implementing distributed rural electrification projects (often in conjunction with international as well as local actors). Examples of centralized governmental programs for distributed rural electrification include programs to diffuse solar systems for community facilities in Brazil (the PRODEEM Programme) and the Township Electrification Programme of the Chinese government (Ku et al. 2003; Goldemberg et al. 2004).

Decentralized/Governmental Organizations

There are numerous examples of community ownership of small scale distributed power generation technologies. These are often supported by NGOs, governments or donors and many focus on electrifying community structures (schools, clinics, etc.) or creating village level grids rather than individual households or private commercial sites. An important aspect of this model is the involvement of the community in decision-making regarding technology choice and implementation strategy. In some cases, this model may be combined with decentralized non-governmental models (see below) in which the community owns the technology but a commercial entity (such as a local agricultural producer) operates the unit and both receive a share of the profits.

Centralized/Non-Governmental Organizations

In this model technology sales (e.g. solar home systems) or distributed electricity service (e.g. fee-for-service or SHS rentals) are by a large commercial actor outside of the rural area. One example of centralized supply of electricity through distributed power generation is the concession system. Concessions provide a company with an exclusive monopoly electricity service territory. In the United States, this means that a company is given exclusive license to provide electricity to a given geographic region with an obligation to serve and abide by regulated tariffs charged to consumers. A modification of this model allows the concessionaire to serve the customers using a mix of technologies that can include both extension of the central grid and the use of off-grid or mini-grid distributed power generation. Due to the higher costs involved in serving rural areas and the lower ability to pay by rural customers, some concession systems are based on bidding for a minimal subsidy from the central government. It is then up to the concessionaire to put together a package of technologies to meet rural electrification needs. Some variant of this model has been used in Argentina, Brazil and South Africa, among others.

Another example of ways in which centralized, non-governmental organizations can play a role in distributed rural electrification is through direct sales or dealer networks. Information is difficult to obtain on the existence of these purely commercial operations. It is likely, however, that these represent only a small portion of the supply in rural areas unless they are through subsidized programs. Examples may include dealer networks for small diesel generators or for small PV systems.

Decentralized/Non-Governmental Organizations

Distributed power systems open up the possibility for small-scale enterprises whose sole purpose is to provide energy technology or services. In this model the locus of activity is with a small local non-governmental actor, which can include non-profit associations, cooperatives, and entrepreneurs. Outside support (financial, technical, and organizational) may exist and be provided, but success or failure depends on the actions of the local actor.

There are numerous examples of these types of small-scale enterprises in both urban and rural areas. In some cases, they provide direct sales of electric power components or systems (e.g. a supplier of small photovoltaic systems). These could be supplied on a cash basis or through financing arrangements. In other cases, they may provide an energy service for a fee. This could include setting up a battery-charging station, renting PV systems, or establishing a micro-utility that serves a village. While these enterprises may be small, in some cases they are linked to a larger organization operating at the state, national or even international level. Soluz Dominica, for example, is majority owned by the American Soluz firm (Allderdice and Rogers 2000; Scheutzlich et al. 2000).

Some actors within this group operate on a purely commercial basis with no subsidies or central support – such as operators of diesel-based mini-grids in Cambodia, vendors of off-the-shelf small PV systems in western China and Kenya. Others have been supported by government and/or private donors and either the entrepreneur or the customer can take advantage of subsidies or other preferential policies. In other cases, a non-profit organization is the locus of the activity, sometimes with an explicit agenda of promoting development and/or environmental protection. In such cases rural electrification can be the main focus of the organization or simply a vehicle to achieve larger social welfare goals (i.e. a health NGO that gets involved in providing electricity for clinic refrigerators). Such organizations can be funded by grants and gifts from both national and international sources. In some cases, the goal of the organization is to develop new models for providing electricity in rural areas, including promoting the development of more commercial enterprises.

Other decentralized non-governmental actors are based on the cooperative principle. These are member-based organizations in which each member has a "share" in the cooperative and its benefits, and participates in the governance of the coop. Any profits from the commercial activity of the cooperative are generally used for improvement of service or returned to the coop members as dividends. These organizations were an important part of the success in rural electrification in the United States.[5] The Philippines is one country where cooperatives have been widely used for rural electrification. In the case of isolated islands, this includes DG-based mini-grids.

Hypotheses

Electricity Service

It is hypothesized that centralized, governmental models will have the largest impact on electricity access levels in the short term as they are specifically designed to reach the largest populations (e.g. large solar home system projects and universal service programs) and have the resources for large-scale projects. Affordability is also highest under centralized programs as these are usually the organizations through which large subsidies (direct or cross-subsidies) are implemented. This is due to the ability of central governments to overcome collective action problems and implement redistributive policies (Shin 2001). Centralized, non-governmental models would have the second highest short-term impact when mandated by the government to meet electrification targets. Their impact may be limited by their need to maintain profitability. Centralized utilities have to balance meeting rural electrification needs with maintaining profitability, creating an inherent tension that can lead to underservice (Foley 1992).

Sustainability

From a financial sustainability perspective, centralized systems benefit from economies of scale, access to conventional financing and donor funds, solving many of the capacity problems of decentralization. However, central control over decentralized technology can entail a number of diseconomies of scale. They may require greater bureaucracy and higher paid employees, and incur higher costs in operations and maintenance if they have centrally-located support personnel (Foley 1992; Shin 2001).

The oft-stated general advantages of political and economic decentralization are the ability to account for local preferences, improved public participation and good governance (Litvack et al. 1998). It is hypothesized that decentralized models are more likely to be institutionally sustainable than centralized models. Improved sustainability is the result of the stake that decentralized actors, particularly those that depend on community support (e.g. local governments or NGOs), have in ensuring well-functioning systems. Decentralized models are also more likely to incorporate local needs, include community members in decision-making processes and obtain community consensus on various aspects of the project (Barnes and Floor 1996;

---

[5]Through preferential policies and a different business model, the coops were able to supply electricity in areas that traditional utilities considered unprofitable. In general, coops in the United States own little generation themselves and are wholesale customers of the larger investor-owned utilities.

Gunaratne 2002). However, one caveat is that it may be necessary to have some involvement by centralized actors that can mobilize resources and have the expertise to undertake necessary repair and maintenance operations (Foley 1992) .

Replicability

Replicability will be highest when the primary organization involved is decentralized. Distributed electrification requires careful consideration of local conditions including technical, social and economic factors (Barnett 1990; ESMAP 2000; Jütting 2003). This is precisely the advantage stressed by the literature on decentralization of government services, including infrastructure (Estache 1995; Shin 2001). By contrast, centralized governmental programs can result in a large number of installations initially, but the ability of the government to continuously call upon state resources for expansion of installations will be limited. Centralized governmental structures are also subject to changing priorities. Similarly, donors also have limited resources and shifting priorities, and large donor programs are not usually replicated over long periods of time in one country.

Technology Choice

First and foremost, the technology must suit the area being electrified in terms of both physical resources (availability of fuels or sunlight) and human resources (availability of trained individuals to install and run systems).

We have focused on two particular technological characteristics. The first is whether the technology is based on the use of a renewable resource or not. This characteristic is important because it reflects resource availability and suitability, as well as larger institutional forces that may favor a particular class of technologies (e.g. donor preferences for renewable energy projects).

The "sustainable development" paradigm has led to a focus on particular renewable technologies for achieving rural electrification such as solar home systems, biomass gasifiers or wind systems (Byrne et al. 1997; van Campen et al. 2000; Biswas et al. 2004). The dispersed and remote nature of rural loads, low consumption, and the difficulties of obtaining fossil fuels can sometimes make renewable energy sources optimal from an electrification perspective, in addition to providing environmental benefits. However, there is a danger that electrification (and by extension, local economic development) becomes an ancillary benefit in an effort to meet environmental goals, particularly those related to global warming. Given the small contribution of rural electricity generation to global warming, more work needs to be done to understand the implications of pursuing environmental and development goals at the same time and how to best align the two objectives.

Technology choices for distributed power, as with centralized power, have both local and global environmental implications. Environmental policies can have impacts on technology choices both through the creation of normative concepts of "clean technologies" and by creating barriers to and incentives for particular technologies. In addition, donor organizations and project NGOs can place an emphasis on a particular technology or class of technologies. Thus, electrification projects may not proceed unless the technology meets certain environmental goals. Programs funded by the Global Environmental Facility are one such example. The GEF has funded a number of rural electrification projects. Due to the GEF's explicit environmental focus, specifically global climate change, these rural electrification projects have used renewable energy technologies (GEF project documents explicitly show how much carbon is saved through renewable electrification rather than diesel generation). Cheaper options, particularly diesel, would not get funded under the GEF program because its priority is global climate change rather than rural electrification.

Often, however, distributed technologies are compared to one another without a comparison of the baseline environmental impact of current energy choices. Thus, a direct comparison of diesel generators and photovoltaics would clearly indicate the environmental advantages of photovoltaics. However, in comparison to the current energy source, diesel could still represent an environmental improvement. This would particularly be true in the case of indoor air pollution resulting from kerosene or candle lighting. If electricity is used for cooking (though this is often not the best option in terms of modern energy sources), then the comparison could be with firewood or other biomass sources or even coal (e.g. in rural china where direct coal burning is common). Depending on their relative costs, it would be possible for diesel in some cases to be the most cost effective technology for reducing the pollutant from the old energy source (on a per kg emission basis).

The second characteristic is whether the technology is deployed through individual units or through a grid. Some technologies are only suitable for a particular mode of operation due to scaling issues (e.g. diesel units have a minimum size making them suitable mainly for micro-grids unless installed at a productive facility requiring higher power). Again, this is important as it reflects resource availability and suitability, including the human resources necessary for either installing individual units or running a small grid (Table 2.3).

Hypotheses

Electricity Service

It is easier to organize the large-scale distribution of individual units (for example, through the donation of solar home systems by national governments or international donors) than the distribution of electricity through a small grid. Even when individuals are purchasing the technology, the individual installations do not

Table 2.3 Examples of technology choices

	Renewable	Non-renewable
Micro grid	Small hydro, wind hybrids	Diesel, coal
Individual installations	Solar home system, pico-hydro	Diesel for productive purposes

require collective action in terms of resource mobilization and decision-making (as in community based projects) that can cause underprovision of public goods (Shin 2001). Individual systems can also often be scaled to meet the differing needs of various segments of the population. A grid system requires an organizational structure that can bill for service and collect payments as well as operate and maintain the system (ESMAP 2000). Conversely, it is easier on a local scale to create mini-grids, particularly ones based on diesel engines. The technology is readily available and familiar and the capital outlays by the end consumer are lower.

Sustainability

There is no hypothesized link between technology choice and financial sustainability. All technologies have an equal potential to perform poorly or well depending on other aspects of the model. This is true as long as proper care has been taken to account for the resource base available. DREMs based on non-renewable technologies do, however, face the problem of shifting fuel costs and thus a model that is financially sustainable at its inception could face problems if diesel or gasoline costs rise sharply.

It is hypothesized that renewables will initially face greater hurdles to sustainability for non-financial reasons. This is primarily due to the relative unfamiliarity of newer renewable energy technologies, resulting in misperceptions regarding maintenance and operational requirements. In markets with more experience in using renewable energy technologies, sustainability should not be affected by this factor (Nieuwenhout et al. 2001; Martinot et al. 2002).

Replicability

For the same reasons that individual installations are likely to result in higher electricity access in the short term, they are also likely to be more replicable. However, there are two caveats to this hypothesis. First, individual installations tend to be renewable energy technologies and, as noted for the sustainability hypothesis, there remains a barrier regarding familiarity with renewable energy technologies and their implementation. Individuals and communities may simply be unaware of renewable energy options, or if they know that renewables are an option, they may not know how or where to purchase such technologies. Second, even if there is knowledge about renewables as an option, there is also a barrier to renewables with regard to human capital. Unlike with renewable energy technologies, there is widespread capability to work with engine technologies for automobiles, mechanical pumps, etc. This pre-existing capability makes diffusion of engines easier (Barnett 1990).

However, small grid systems can result in higher quality service over the long term for populations that are not too highly dispersed. If the organizational and financial obstacles issues can be resolved, it is easier to supply larger populations through the grid than through individual units. Financing for a micro-grid can be easier to obtain than credit for an individual installation, making it possible to fold capital costs into the monthly fee and reduce up-front costs to the end-user, thereby increasing affordability. The ability to have multiple sources of supply in even a small grid can also result in higher reliability and service quality (ESMAP 2000; Zerriffi 2007; Banerjee 2006).

Target Customers

The nature of the customer base for distributed rural electricity has a direct impact on electricity access and financial performance. End uses can be roughly divided into: household electrification, productive uses (often agriculture) and community structures (clinics, schools, street lighting).

From a technical perspective, the nature of the demand depends on the types of customers and their uses of electricity. This includes the level of peak demand, the hours electricity is used, the types of appliances run and the shape of the demand curve. This can also have an effect on the relative economics of supplying electricity via the grid versus distributed systems (ESMAP 2000). In some cases, rural electricity use is highly concentrated in just a few hours of the day, which can exacerbate the peak load problems of centralized utilities and drive up the cost of centralized rural electrification (Howells et al. 2006).

The willingness and ability to pay for electricity also varies according to the nature of the target customers. Wide variations in household income will affect willingness and ability to pay. Economic inequalities that can affect both willingness and ability to pay can be seen not only between countries or regions of countries, but even within a single community or closely neighboring communities (Elbers et al. 2004). The degree to which electrification is limited to higher income households or targets all households, regardless of income, will have an impact on project performance.

In general, commercial enterprises (including agricultural users) that use electricity to generate income have a greater ability and willingness to pay for electricity than households. There are numerous productive activities (e.g. agricultural producers, small shops, and textile workers) in rural areas that can benefit from increased access to electricity and provide a solid customer base with the potential to recover more of the costs of providing electricity. Electricity can be used either to improve the production process (increasing throughput or providing consistent quality products by replacing hand-powered machines with electric ones) or indirectly by changing work conditions (allowing a shop to stay open later or a textile worker to work in the evenings) (Etcheverry 2003; Fishbein 2003; Meadows et al. 2003).

These commercial customers can also act as an anchor consumer for the electrification project, allowing more of the capital costs to be recovered. For example, an agricultural producer can use the electricity during the day to process agricultural produce and in the evenings the unit can supply a basic amount of electricity (enough for lights or perhaps a radio) to domestic users. For enterprises that install their own generating unit, they can potentially earn additional income by supplying the community with electricity when not using the power for their primary activity.[6]

---

[6]This will be true as long as the tariffs can exceed the operating costs of the unit, which will depend upon the customers' willingness and ability to pay.

Table 2.4 Characteristics of different target customers

	Households	Productive activities	Community structures
Nature of demand	Low per household, at peak times	Low to high depending on activity, more constant	Low/medium
Ability to pay	Varied but mostly low	Higher	Low
Willingness to pay	Varied but can be high	High	Low
Example observations	Solar home program in Bahia, Brazil	Early SHP for agriculture in China	NGO projects for clinics/schools in Cambodia

The customer base not only affects willingness and ability to pay, but also the impact of electrification on welfare. Electrification that reaches into the poorest households in rural areas will have very different social welfare impacts than electrification efforts that only reach richer rural households. Similarly, electrification that includes community buildings or commercial uses will have different welfare impacts than if the target customer base is solely households (Table 2.4).

Hypotheses

Electricity Service

Improvements in electricity access over the short term will be higher when the target customer base is rural households due to the higher number of rural households as compared to productive enterprises or community structures. However, access levels can be even higher if there is a combination of target customers. Provision of electricity to a combination of households and productive activities improves the financial performance and thereby can increase the number of households electrified for a given investment (ESMAP 2000; Martinot et al. 2002). Supply reliability and quality should also be higher when productive activity is included in the target customer base due to higher reliability and quality power demands of most productive activities (Allderdice and Rogers 2000; van Campen et al. 2000; Meadows et al. 2003).

Sustainability

Financial sustainability will likely be better when electricity supply accounts for the needs of rural productive enterprises because productive enterprises have a greater ability to pay for power (ESMAP 2001). In comparison to households where energy expenditures tend to be kept below 15%, energy expenditures by productive activities can range widely, in some cases reaching over 50% of production costs (Meadows et al. 2003). Furthermore, improving the productivity of rural enterprises is believed to have additional development benefits that can change household incomes, resulting in better financial performance over the long-term for household electricity supply (Allderdice and Rogers 2000; ESMAP 2000; Meadows et al. 2003). Financial sustainability is worst when the target customer is only community structures (schools, clinics, etc.). The collective action problem of having community members pay for such electrification can be large. A middle ground may be village-level programs in which a rural enterprise acts as an anchor consumer. This improves the economics since the anchor consumer can better afford the electricity used for their productive activity while still providing electricity to rural households and community structures at a reduced rate.

Institutional sustainability will also be highest when productive activities are included in the customer base. The commercial and productive users of the electricity have a strong vested interest in continued service in order to maintain higher productivity and incomes. This should result in strong institutional support for the project by those users, helping to overcome barriers that arise. At the other end of the scale are individual households. While these households would also have an interest in continued service, they face a collective action problem and any individual household (with the possible exception of local influential elites) will have less influence than a local agricultural producer or other productive activity.

Replicability

Serving productive activities should increase the replicability of distributed electrification models from a financial perspective. The reasoning is the same as for securing financial sustainability of the model; productive users are able and willing to pay more for electricity.

From an institutional and organizational perspective, however, the relationship between the target customer base and replicability is complicated and may not always be consistent. One factor is the collective action problems that can make action at the local level difficult (Shin 2001). Serving either households or productive activities alone, rather than as part of a community electrification plan, reduces the need to overcome collective action problems. A second issue is the need to account for the local context. Productive activities can range widely and are highly heterogeneous in terms of the amount of electricity required, the importance of electrification for business and willingness to pay for electricity (Meadows et al. 2003).

Financial Structure

There are a number of aspects of the financial structure of distributed rural electrification efforts that can affect electricity service, sustainability and replicability. Two aspects of the financial structure are emphasized in this study due to their direct relationship with the outcomes. The first is the financing of capital costs and the presence of capital subsidies or other support mechanisms. The second is the recovery of operating expenses and the tariff structure.

In a discussion of the financial structure of a business model for distributed rural electrification, there is a temptation to try to characterize models according to whether they are commercially viable or not. However, the distinction between commercial and non-commercial models is somewhat difficult in this context, given the system of subsidies often in place to support electricity access. While there are examples of purely commercial operations, many operations that might be classified as commercial for the purposes of this study take advantage of such subsidies or other preferential policies and do not have true cost-recovery or profitability. Even when an actor is considered a commercial entity, they may not actually depend on cost recovery from consumers in order to continue operations. Thus it is necessary to distinguish situations in which a commercial enterprise is involved but their compensation is not tied to the profitability of the rural electrification. For example, a company that receives a contract to install 10,000 solar home systems as part of a large donor/government program may be responsible for planning and implementation, but its revenue is the result of a government subsidized contract rather than commercial sales or service to end-users. We have, therefore, not used this distinction in our analysis, instead relying on understanding the underlying nature of the financial structure of the model.

It is tempting to include a categorization of organizations according to whether they could be considered commercial or non-commercial. However, as will be discussed below, this is not necessarily a useful distinction and the characteristics usually associated with commercial versus non-commercial activity are best served by separate examination under the variable of "financial structures."

Capital Financing: Due to continued low diffusion rates and the relatively high costs of the various technology options (particularly renewables), significant state government involvement has often been considered necessary for covering capital costs. This capital financing support can come in the form of technology installations, loan guarantees with favorable terms, or partial subsidization. Capital financing can also be undertaken by private lenders, micro-finance organizations, and international donors. Important factors determining the nature and amount of capital financing required include the degree to which tariffs are used to pay back capital and the degree to which end-users have to provide equity.

Recovery of Operating Expenses: All systems require long term maintenance, without which systems can fail significantly short of their design lifetimes. The financial structure must account for recovering these operating expenses, including possible large investments in equipment replacement over the long term. In some cases, operating expenses are entirely recovered through the tariff for electricity sales. However, as with capital expenses, significant subsidies often exist for operating expenses and are part of the financial structure of the enterprise. In other cases, operating expenses are not part of the financial structure at all and individuals are responsible for long-term maintenance (for example in the PV cash market) or eventual replacement.

Table 2.5 summarizes the different options available for financing of both capital and operational costs and provides examples of how different distributed rural electrification models can be financially structured. A number of DREM have multiple options for

Table 2.5 Example DREMs for different financial structures

	Grants/subventions	Cost recovery tariff	Cross-subsidy
Capital costs	Technology donations, utility based systems, micro-grid entrepreneurs	Technology cash markets, micro-grid entrepreneurs, utility based systems	Utility based systems
Operational costs	Government programs or contracts	Technology cash markets, micro-grid entrepreneurs, utility based systems	Utility based systems, government tax transfers

covering both capital and operational costs. For example, utility-based systems that provide electric power to customers in the most remote areas of their concession through distributed generation can, in theory, cover such costs through government grants, a full cost-recovery tariff to the rural customer or through a cross-subsidy from their urban or industrial customers. This will depend on a number of factors, including government policies and regulations. On the other hand, some DREM can only cover capital or operating costs through one or two of these methods. For example, micro-grid entrepreneurs are unlikely to have the mix of customers necessary to cross-subsidize their poorest consumers with higher tariffs on richer consumers.

There are different ways in which to the financial structure of the model can be organized. In general, however, the business models can be divided into those that emphasize direct sales of technologies in rural areas (such as for small solar home systems) and those that emphasize the provision of service in rural areas (and therefore act like a traditional utility, charging a tariff and maintaining system ownership). Sales and service models have different cash-flow characteristics and require different financial inputs by the end-use customers, ultimately impacting the affordability of the technology or service and the financial viability of the distributed power source. Under the sales model, individual consumers (households, commercial enterprises, government buildings, etc.) purchase and install a unit on their own site. These can be straight cash markets (e.g. solar home systems in China and Kenya) or have some form of end-use customer financing (e.g. the Enersol program in the Dominican Republic). The second model provides the electricity through fee-for-service arrangements or provides the technology under a rental agreement. This includes micro-grid installations, leasing of solar home systems (where ownership is not transferred to the user), utility ownership and PV-based battery charging stations (e.g. the IDEEAS program for SHS leasing in Brazil).

Hypotheses

Electricity Service

The relationship between the financial structure of the model and short-term changes in electricity service (particularly access) is complicated. If the financial structure is considered in terms of the level of government or donor subsidies, then electricity access will differ depending on whether subsidies are low, medium or high. If subsidies are low, electricity access will be restricted to those that can afford to pay the full market cost of access.[7] At medium levels of subsidies costs are shared between governments and end-users and, in the short term, electricity access rates could be higher as government funds can stretch further with consumer burden-sharing. This will be true as long as government budgets can support the outlay. At the limit, however, when technologies are given away (i.e. installation is 100% subsidized), electrification rates could be lower. Limited government budgets make it unlikely that all demand could be met, but incentives for individuals to invest would be lower based on the belief that free installation is a right and that the best course of action is to wait for the government to install the electricity supply for free (Martinot et al. 2002). Furthermore, high subsidies have been shown to damage viable markets, reducing long-term access (Martinot et al. 2002). Table 2.6 shows the hypothesized impacts of these different subsidy levels on the end-user (in terms of cost and their incentives to invest their own capital) and on changes in electricity access (over the short and long term).

A major barrier for rural populations is the high up-front cost of electrification. This applies equally to obtaining a grid connection (when unsubsidized) and purchasing a photovoltaic system for the home. Any capital financing model that allows end-users to spread these up-front costs out over time (e.g. service models that fold the costs into the monthly charge or sales models that incorporate customer credit schemes) will increase affordability and access. Service models are also hypothesized to increase affordability and access due to the possibility of cross-subsidies with other users (e.g. the anchor consumer, discussed above).

Sustainability

Financial sustainability will be greater when the financial structure of the model emphasizes full recovery of operating expenses and at least some recovery of capital

Table 2.6 Hypothesized impact of subsidy levels on end-users and changes in access

	Low subsidies	Medium subsidies	High subsidies
Cost to end-user	High	Medium	Low
Incentives for end-user	Medium-low	High	Low
Short-term access	Low	Medium	High
Long-term access	Low-medium	Medium	Low

Note: Astute readers will notice that there is no level of subsidy resulting in "High" levels of access over the long-term. While some combination of low and medium targeted subsidies is hypothesized to be best in delivering electricity over the long-term, the lack of a "High" designation is meant to reflect the magnitude of the remaining challenge in serving rural populations, particularly in South Asia and Africa.

---

[7]High costs could also act to limit consumption by some customers rather than completely eliminate access. This is reflected in the cash PV markets where PV cell sizes are often very small and users buy additional modules as their incomes allow.

(Barnes and Floor 1996). Obviously, complete cost recovery through tariffs is the most sustainable. Cross-subsidies can be sustainable and the reduction in economic efficiency justified as long as they do not place a large burden on the subsidizing customers or the utility and meet certain requirements for efficiency (Barnes and Floor 1996; Beato 2000). Beato outlines different cross-subsidization schemes and shows how only some will be sustainable and maintain efficiency. In the case of cross-subsidies based on average costs, they can even increase total welfare (Beato 2000). Grants tend to be unsustainable as they come from state budgets, which have to be negotiated annually, or from donations which are often short-term.

The financial structure of the DREM also has an impact on non-financial aspects of sustainability of the model. In particular, service models (both micro-grids as well as service models based on individual installations) are hypothesized to be more sustainable than cash sale models. Service models have a built-in provision for maintaining the system whereas many cash sales are not accompanied by maintenance and replacement provisions.

Replicability

Financial structures that are highly dependent upon external capital financing, such as donor aid, are difficult to replicate. Removal of donor aid has lead in the past to a collapse of the system of entrepreneurs (Martinot et al. 2002). This is consistent with the general literature on fiscal decentralization, which emphasizes balance between decentralization of revenues and expenditures (Bird 1994). Financial replicability is therefore higher when subsidies are low and when capital costs are recovered since there is no dependence on state budgets, donations or other customer classes. However, there is a limit to replicability under those circumstances if the costs are higher than the ability or willingness to pay. In order to serve those at the lowest income levels, models with some subsidies for both capital and operational costs may be more replicable than financial structures that require full cost-recovery. Replicability will also be higher when the financial structure of the DREM results in increased income or savings to the technology adopter (Barnett 1990). Another option for overcoming first-cost problems is to spread the costs over time and therefore capital financing models that incorporate this feature (either through service or consumer credit) will also be more replicable.

There is no hypothesized relationship between the financial structure of the DREM and institutional (i.e. non-financial) aspects of replicability. The more important consideration is that the financial structure of a DREM being replicated fits within the new institutional context. Therefore, a financial structure that requires monthly tariffs to be paid and was successful in a setting where households have regular wage labor remuneration may have to be adapted in a setting where household income is closely tied to crop harvesting and regular monthly income is replaced by a large yearly or semi-yearly cash inflow. Similarly, adjustments in either the DREM or institutional setting would have to be made if a DREM based on extending credit to households is transferred to a setting in which rural household credit facilities don't exist and lending institutions are geared primarily to urban and commercial customers.

Summary of Hypotheses

In general, most hypotheses indicate improved long-term performance from non-governmental or private sector organizations focused, at least in part, on improving electricity service to productive enterprises. Similarly, an emphasis on cost recovery for both capital and operational costs should improve the long-term sustainability and replicability of distributed rural electrification models. However, in the short term, improved service for a larger portion of the population is likely to be the result of financially non-sustainable and replicable programs aimed at households rather than productive activities and organized through centralized governmental organizations or large donor programs (Table 2.7).

Tradeoffs: It is clear that some institutional models are hypothesized to perform better on some metrics of rural electrification and worse on others. It is hypothesized that the tradeoffs of using one model over another may be minimized by the use of hybrid models that combine the benefits of governmental and non-governmental actors and/or centralized and decentralized actors. In particular, models that combine community-based electrification with decentralized commercial actors may achieve higher levels of electricity access and be more sustainable and replicable while also improving cost recovery. One means to accomplish this may be through the use of anchor consumers in the community who would use the electricity for productive commercial activities.

Control Variables

As data was collected on the different cases and categorized into the dependent and independent variables, some of the relevant information did not fit into the existing variables. These data fell into two basic categories. First was data relevant to the institutional context. The presence of subsidies for either capital costs or operating costs makes a large difference regarding the viability of a distributed electrification model. It improves the finances of the model (as long as the subsidy is sustainable) and makes other models less competitive. In addition to subsidies, there was a need to categorize the policy and regulatory regimes more generally in order to capture the impact of the institutional context on the distributed electrification models. Control variables were therefore created for these four factors in which the level of capital and operating subsidies were given a low-medium-high scale and the policy and regulatory regimes were characterized as favorable, neutral or unfavorable.

Second, was data relevant to the physical context, particularly whether the population was located in remote regions and the density of the population. Remoteness bears directly on the potential viability of grid extension and on the potential difficulties related to project management and operations and maintenance when these are more centrally controlled. The density of the population is relevant for the relative viability of the grid, micro-grids and individual installations. These two factors were therefore included as control variables in the study.

Table 2.7 Summary of hypotheses

	Organizational form	Technology choice	Target customers	Financial structure
Changes in electricity service	Centralized, particularly governmental, programs will result in the greatest *short-term* increase in electricity access for rural households	Programs for individual installations (e.g. solar home systems) will result in the greatest *short-term* increase in electricity access	Programs that include electrification of productive activities will have higher levels of access and service quality	Programs with high subsidies will have the highest short-term impact on access
Sustainability	Decentralized organizations will be more sustainable than centralized models, particularly when accounting for institutional factors important for sustainability	Renewable technologies have the potential to be more financially sustainable due to being immune to fuel price variations. Non-renewable technologies are more sustainable institutionally due to the unfamiliarity of renewable technologies and lack of associated widespread maintenance mechanisms.	Sustainability (both financial and institutional) will be higher when productive activities are included	Models with full cost-recovery (particularly of operating expenses) will be the most sustainable financially
Replicability	Decentralized models will be more replicable than centralized models	Individual installations will be more replicable than mini-grids. Non-renewable mini-grids will be more replicable than renewable based mini-grids	Models that emphasize individual users (productive users or households) will be more replicable than those that emphasize groups of users. Models that emphasize productive activities will be more replicable than those that only focus on households	Models that include full recovery of capital expenses will be the most easily replicable, particularly when the model includes some method for financing capital, but are limited by the willingness and ability to pay of consumers. Models with modest and targeted subsidies will be the most replicable overall.

There are a number of factors that influence how the distributed rural electrification models impact the immediate outcomes. These effectively act as control variables, influencing outcomes in a way that creates variation, even if choices made for the DREM are identical.

Remoteness of Customers: The more remote the customers are, the less likely that centralized technologies can be used to meet their electricity needs. This makes distributed electrification, even if implemented by centralized organizations, more viable.

Density of Customers[8]: The density of customers has an impact on whether or not mini-grid based distributed electrification is a viable alternative or even the best alternative. Mini-grid electrification, as opposed to the installation of single units at the end-users home or business can be more cost-effective and provide higher levels of service if the population density is high enough. On the other hand, mini-grid models often require different types of financial models due to higher capital costs and increased need for an on-going operations and maintenance tariff.

Capital and Operating Cost Subsidies: The impact of the subsidy structure is discussed above as part of the financial structure of the Distributed Rural Electrification Model. It is a combination of the direct capital and operating cost recovery of the DREM along with the presence of capital and operating cost subsidies that determine the sustainability and replicability of the DREM.

Policy Regime: The policy regime in place may be favorable, neutral or unfavorable when it comes to distributed electrification. Policies that encourage local development through infrastructure enhancements or that provide for reduced taxes, access to loans or technical support are all favorable for distributed electrification. Policies that privilege expansion of centralized grids or that increase transaction costs for distributed electrification efforts are unfavorable.

Regulatory Regime: The regulatory regime in place may be favorable, neutral or unfavorable when it comes to different types of distributed electrification. Broadly favorable regulatory regimes would have rules and procedures that are suited for the types of organizations and technologies involved in distributed electrification. Substantively, a favorable regime would provide for access to the grid to sell surplus electricity, tariff structures that don't penalize distributed generation suppliers, and utility service territories that don't prevent more appropriate distributed solutions from operating.[9] Neutral regulatory regimes do not support distributed electrification efforts, but they also don't have regulations in place that hamper distributed electrification. Unfavorable regimes make it difficult for distributed rural electrification systems to be installed and operated legally or create conditions that privilege centralized options or make it impossible for distributed systems to be viable. In addition, the regulatory regime might be favorable or unfavorable for a particular type of

---

[8]It is possible to have a population that is remote from major urban centers and centralized power grids but is, at the same time, relatively closely spaced.

[9]This study looks at both "regulatory governance" and "regulatory substance" in determining whether the regulatory system is favorable, neutral or unfavorable towards distributed electrification models (Reiche et al. 2006).

distributed electrification model. For example, some types of organizations (e.g. large utilities or local governments) may get preferential treatment or access to resources in implementing distributed electrification efforts in comparison to other types of organizations (e.g. cooperatives or local entrepreneurs).

Case Selection

The effect of the Distributed Rural Electrification Model (DREM) on the outcomes of rural electrification efforts is analyzed through a case-study approach. In addition, we use multiple cases per country to account for the country-specific institutional factors that play a role in both choice and performance.

In order to compare different institutional models for the use of distributed power in rural areas, it is desirable to have variation on two dimensions. First, there should be variance within a country in the models used for distributed rural electrification. These comparisons within countries will be used to determine how different characteristics of the distributed rural electrification models impact electricity service, sustainability and replicability. By making the country the first unit of analysis, it is possible to control for the institutional and technological factors such as the regulatory system, resource availability, and technology availability.

Second, it is desirable to have variance across institutional and technological factors. Comparisons across countries will be used to determine the effect of those country-level factors and whether the results hold across a range of regulatory, policy, and technical environments.

A two-stage case selection process was followed. In the first stage countries were selected according to three criteria. First, the countries must have had significant experience in distributed rural electrification. Countries that have not progressed beyond the demonstration phase or limited implementation do not provide the necessary historical experience to judge performance.[10] Second, the countries must have experience with different distributed rural electrification models.[11] Third, there had to be variation among the countries in their institutional, regulatory and policy regimes governing rural electrification.

In the second stage, individual observations were chosen within each country case study. These observations (of particular distributed rural electrification models,

---

[10] It is important to recall that this study is of the performance of business models for distributed rural electrification. Therefore, limiting the universe of cases to those countries that have experience with distributed rural electrification does not raise concerns regarding case selection bias or of selecting cases based on values of the dependent variable. If, instead, this was a study of the presence of distributed rural electrification, then this would obviously be a concern.

[11] This is the reason that there are no African countries included among the final list of countries. While there have been some successes (e.g. PV in Kenya), generally speaking the experience has been much more limited and does not provide the basis for making conclusions. However, some of the conclusions of this study could be applied to future efforts in Africa.

and to the extent possible, of individual installations using that model) allow us to move from a small-N study of a few countries to a medium-N study.

There are, of course, many countries in which decentralized options for power supply in rural areas are being implemented. One only has to look at the list of Global Environment Facility projects to see how widespread the trend is to use (mainly renewable) distributed technologies. This is generally due to the remote nature of rural populations and the poor financial or technical performance of many grid systems, making off-grid or mini-grid options attractive. However, while the trend is broad, there has not been a deep experience with distributed rural electrification in most countries. That is, while many countries have experimented with using off-grid or decentralized options, the experience has often been limited in both technological and organizational scope as well as in terms of the magnitude of the effort.

A short list of countries was chosen based on the fact that they had the desired variation as described above. This short list was based on a review of the literature and interviews with experts having a range of experience across a number of countries. One area in which expert information has been particularly valuable is in understanding the importance of diesel generators for rural electrification. There is little information in the written literature on diesel generators, which generally operate on a completely informal basis outside the scope of the regulatory system. This is known to occur in a number of countries, but the extent is often unknown and the particular commercial arrangements are also generally unknown.

From this short list, three countries were chosen for detailed study based on the factors listed above. Those three countries are Brazil, Cambodia and China. The following sections outline some of the important characteristics of distributed rural electrification in these three country case studies. Each of these countries also has the characteristic that a large part of the distributed rural electrification effort has been through a single dominant model (that is different in each case). In addition to this dominant model, there are alternative models in each country. The presence of a dominant model and alternative models allows us to make comparisons within each country while also maximizing scarce resources by being able to focus primarily on the dominant model. Table 2.8 summarizes the within country variations in each country for each of the three independent variables. In addition, secondary literature on other cases may be used as supporting evidence (or as counter-evidence) as appropriate.

As a result, the observations used in this study vary in both time and scope. In some case, observations are of DREMs that were implemented decades ago (e.g. some small hydro plants in China). In other cases, the DREM was only implemented recently (e.g. the Township Electrification Program in China). Similarly, some observations are of programs (again, the TEP in China) while others are of individual installations. In some cases, multiple observations may be made that are not completely independent. As noted in King and Keohane, it is possible to increase the number of observations by looking at sub-units of an observed case or by looking at different time periods. Examining sub-units within an observational unit (e.g. particular installations under an electrification program) is a valid method for increasing the number of observations as long as the hypotheses

Table 2.8 Variation of cases on independent variables

	Brazil (six Cases)	Cambodia (six Cases)	China (six Cases)
Organizational form	*Dominant*: Centralized utilities *Alternative*: Coops, NGOs, small entrepreneurs	*Dominant*: Small entrepreneurs *Alternative*: Government and international donor projects	*Dominant*: Local governmental and private, some hybrid/dual *Alternative*: Decentralized private tech dealers, centralized governmental
Technology choice	*Dominant*: Diesel *Alternative*: Biomass, PV	*Dominant*: Diesel *Alternative*: Biomass, PV, small hydro	*Dominant*: Small hydropower *Alternative*: Small thermal, PV, wind
Target customer base	*Dominant*: Households *Alternative*: Varied	*Dominant*: Village electrification *Alternative*: Households	*Dominant*: Village and higher electrification *Alternative*: Individual systems
Financial structure	*Dominant*: Subsidized connections and low income consumers *Alternative*: Market prices with cost recovery	*Dominant*: Market prices *Alternative*: Highly subsidized	*Dominant*: Cost-plus regulated prices *Alternative*: Subsidized, cash markets

developed apply to the new observations. Similarly, multiple observations over time can be made of the same case. However, in doing so, there is no longer complete independence between observations of the dependent variable and so more limited conclusions must be drawn when using these types of observations. To the degree possible, the use of such observations will be avoided due to this limitation (King et al. 1994).

As with any study of this nature, case selection posed a problem. It is not feasible to gather data on every single distributed generation initiative or project ever installed in the country.

This leads to three potential sources of bias in the results:

1. Lack of information on older projects, particularly those that have failed. It is, of course, difficult to observe what is no longer there and if it wasn't adequately documented in secondary literature, it is difficult to assess. In some cases, it was possible to obtain limited information on these efforts through interviews. In those instances, they were not treated as full cases for the study, but this information was used to help support general conclusions drawn from the cases.
2. Lack of information regarding smaller and less public efforts, such as independent diesel generators in the Amazon. Similar efforts to fill in some information about these distributed models were pursued as with the first case.

3. In some cases, detailed information came only from the parties responsible for a particular electrification model. This could lead to potential bias in some of the results, though in all of these cases, there were both negative and positive assessments provided, indicating that there was no systematic bias towards presenting the information in an overly positive light.

While it would have been ideal to include the entire universe of cases, what is important for this study is that there was no systematic bias in the cases selected since there was no a priori rejection of cases on the basis of technology, organization, etc.

References

Acker RH, Kammen DM (1996) The quiet (energy) revolution: analysing the dissemination of photovoltaic power systems in Kenya. Energy Policy 24(1):81–111

Allderdice A, Rogers JH (2000) Renewable energy for microenterprise. National Renewable Energy Laboratory, Golden, pp 1–70

Banerjee R (2006) Comparison of options for distributed generation in India. Energy Policy 34:101–111

Barnes DF, Floor WM (1996) Rural energy in developing countries: a challenge for economic development. Annu Rev Energy Env 21:497–530

Barnett A (1990) The diffusion of energy technology in the rural areas of developing countries: a synthesis of recent experience. World Dev 18(4):539–553

Beato P (2000) Cross subsidies in public services: some issues. In: Sustainable development department technical paper series. IFM-122. Inter-American Development Bank, Washington, DC, p 25

Bird R (1994) Decentralizing infrastructure: for good or ill? WP 1258. The World Bank, Washington, DC, p 34

Biswas WK, Diesendorf M et al (2004) Can photovoltaic technologies help attain sustainable rural development in Bangladesh? Energy Policy 32:1199–1207

Byrne J, Shen B et al (1997) The economics of sustainable energy for rural development: a study of renewable energy in rural China. Energy Policy 26(1):44–54

Cabraal RA, Barnes DF et al (2005) Productive uses of energy for rural development. Ann Rev Env Resour 30:117–144

Elbers C, Lanjouw PF et al (2004) On the unequal inequality of poor communities. World Bank Econ Rev 18(3):401–421

Elias RJ, Victor DG (2005) Energy transitions in developing countries: a review of concepts and literature. PESD Working Papers. #40. Program on Energy and Sustainable Development, Stanford University, Stanford, CA, p 38

ESMAP (2000) Mini-grid design manual. 21364 (ESMAP Technical Paper 007). Energy Sector Management Assistance Programme, Washington, DC

ESMAP (2001). Best practice manual: promoting decentralized electrification investment. ERM248. Energy Sector Management Assistance Programme.

Estache A (ed) (1995) Decentralizing infrastructure: advantages and limitations. World Bank Discussion Papers. The World Bank, Washington, DC

Etcheverry J (2003) Renewable energy for productive uses: strategies to enhance environmental protection and the quality of rural life.. Department of Geography and Institute for Environmental Studies, University of Toronto, Toronto, p 49

Fishbein RE (2003) Survey of productive uses of electricity in rural areas. Africa Energy Unit, World Bank, Washington, DC, pp 1–50

Foley G (1992) Rural electrification: the institutional dimension. Utilities Policy 2(4):283–289

Goldemberg J, Rovere LL et al (2004) Expanding access to electricity in Brazil. Energy Sustain Dev VIII(4):86–94

Gunaratne L (2002) Rural energy services best practices. Nexant Sari/Energy for United States Agency for International Development, Washington, DC, p 77

Howells M, Victor DG et al (2006) Beyond free electricity: the costs of electric cooking in poor households and a market-friendly alternative. Energy Policy 34(17):3351–3358

Jütting J (2003) Institutions and development: a critical review. Working Paper No. 210. Organization for Economic Cooperation and Development (OECD), Development Centre, Paris, p 45

King G, Keohane RO et al (1994) Designing social inquiry: scientific inference in qualitative research. Princeton University Press, Princeton, NJ

Ku J, Lew D et al (2003) Sending electricity to townships: China's large-scale renewables programme brings power to a million people. Renewable Energy World, pp 56–67.

Litvack J, Ahmad J et al (1998) Rethinking decentralization in developing countries. Sector Studies Series 21491. World Bank, p 52.

Martinot E, Chaurey A et al (2002) Renewable energy markets in developing countries. Annu Rev Energy Env 27:309–348

Meadows K, Riley C et al (2003) A literature review into the linkages between modern energy and micro-enterprise. ED03493. AEA Technology plc for the UK Department for International Development, Oxfordshire, p 33.

Nieuwenhout FDJ, van Dijk A et al (2001) Experience with solar home systems in developing countries: a review. Prog Photovoltaics Res Appl 9:455–474

Pachauri S, Spreng D (2004) Energy use and energy access in relation to poverty. Economic and Political Weekly, pp 271–278.

Ramani KV, Heijndermans E (2003) Energy, poverty and gender: a synthesis. International Bank for Reconstruction and Development, Washington, DC

Reiche K, Tenenbaum B et al (2006) Promoting electrification: regulatory principles and a model law. Energy and Mining Sector Board Discussion Paper No. 18. Energy and Mining Sector Board, World Bank, Washington, DC, p 49

Scheutzlich T, Klinghammer W et al (2000) Financing of solar home systems in developing countries: the role of financing in the dissemination process. Eschborn, Environmental Management, Water, Energy, Transport Division, Deutsche Gesellschaft für Technische Zusammenarbeit (GTZ) GmbH, p 52

Shin R (2001) Strategies for economic development under decentralization: a transformation of the political economy. Int J Public Adm 24(10):1083–1102

Tong J (2004) Small hydro power: China's practice. China WaterPower Press, Beijing

Tongia R (2003) The political economy of indian power sector reforms. PESD Working Papers. PESD Working Paper #4 (Revised). Program on Energy and Sustainable Development, Stanford University, Stanford, CA, p 76

van Campen B, Guidi D et al (2000) Solar photovoltaics for sustainable agriculture and rural development. #2. Food and Agriculture Organization, United Nations, Rome, pp 1–76

Zerriffi H, Dowlatabadi H et al (2007) Incorporating stress in electric power systems reliability models. Energy policy 35:61–75

Chapter 3
Distributed Rural Electrification in Brazil

Keywords Brazil • Monopolies • Regulation • Subsidies • Utilities

Introduction

At the time of this research approximately 8 million people were without electricity in Brazil. This included roughly 1.2% of urban households in 2002, but the bulk lived in rural regions, particularly in the Amazon region and the Northeast (up to 60% of rural households in some regions with a national average of 27% of rural households without electricity in 2002) (ESMAP 2005). A more recent estimate shows over 4 million still do not have electricity access (IEA 2009). It is difficult to see how Brazil, can accomplish its goals of universal service without resorting to some form of decentralized electricity production. At the same time, distributed solutions face a number of institutional barriers that have either precluded their implementation or created conditions unfavorable to their success, even when they may be the best solution from a purely technical perspective.

DG technologies ranging from diesel generators to solar home systems already have a long history in Brazil. Currently, distributed electrification in Brazil is dominated by the centralized utilities installing and operating distributed technologies in order to meet their regulated service mandates. This chapter first examines the successes and failures of distributed rural electrification efforts in Brazil including a number of activities that have been outside of the utility system.

The Institutional Context for Distributed Electrification in Brazil

The use of distributed power generation in Brazil for rural electrification has to first be put in the larger context of the structure of the Brazilian electric power sector and recent government programs and imperatives. The electricity industry in Brazil

Information on Distributed Rural Electrification Models in Brazil comes from a combination of secondary sources and primary interviews conducted in Brazil (September 2005) and Washington DC (2005 and 2006)

has undergone a series of significant institutional changes over the last century. A comprehensive review of this history is beyond the scope of this chapter and has been covered by others (de Oliveira 2003). However, it is worth noting that in the beginning it was the private sector that built, owned and operated the electricity system. By 1950 the installed generation was roughly equally owned by private and state interests. However, private ownership was static from 1935 until the most recent era of reforms in the mid-1990s while the state sector continued to grow. This shift from a private sector to state sector system was less the result of outright nationalization as the result of fiscal and other policies by the Brazilian government that made investment in the electricity industry unattractive for private interests (de Oliveira 2003).

By the time of the latest reform era, the Brazilian system consisted primarily of federally owned electricity assets, particularly the large hydropower plants and the transmission system, and state-level government utilities. The federal government owned a little over half of the generation (with the rest primarily in the hands of the states) but less than a quarter of the distribution was through federally owned utilities (while state utilities distributed roughly three quarters of the electricity). Eletrobras was established as the federal holding company for electricity assets and the federal government divided the country into four regional suppliers (Chesf, Furnas, Eletrosul and Eletronorte). The system was regulated formally by the National Water and Electrical Power Department (DNAEE), though the finance ministry exercised a significant power over tariffs (for example, to ensure macro-economic stability). As noted above, tariffs were insufficient to recover investment costs. In order to maintain the legally mandated rates of return on investment, the difference was put into their balance sheet as the Conta de Resultados a Compensar (CRC), an amount that in theory could be recovered by tariffs at a later time (de Oliveira 2003).

However, international trends in the electricity sector towards reform and the imbalance between investment costs and revenues led to a period of restructuring starting in the nineties. Some of the distribution utilities were privatized, independent power producers were encouraged, and an independent regulator was established called the Agência Nacional de Energia Elétrica (ANEEL). While some of the utilities were privatized, this did not change the highly centralized nature of the Brazilian electricity system. ANEEL oversees a system of monopoly service territories granted on a concession basis, much like in the United States. Concessionaires have rights (exclusivity in their service territory) and obligations (universal service and regulated tariffs) (Br.Ind September 2005; Br.ANEEL September 2005) (Table 3.1).

At the same time, there was renewed interest in rural electrification for equity reasons, resulting in a number of new programs. Enshrined in the Brazilian constitution of 1988 are guarantees for basic needs and social solidarity and requirements for the government to provide, directly or indirectly, public services. It has also been argued that since electricity is required to meet those basic needs, it should also be considered as part of the constitutionally guaranteed services provided poorer populations in Brazil (Paz et al. 2007). ANEEL established tariffs that reflect the notion that poorer populations (both rural and urban) should pay reduced tariffs (discussed further below). The tariff structure requires concessionaires to cross-subsidize their low-consuming customers (on the assumption that low consumption is correlated with low income) with higher tariffs for other consumers. In addition, special funds were established (even before the reform period and the latest constitution) to reduce

Table 3.1 Example tariffs from Minas Gerais

Convencional	Resolução n° 310, 06/04/2006		
	Tarifa isenta de icms r$	Tarifa sujeita a icms r$	% ICMS
B1 – Residencial Normal			
Independente do onsume	0,427964	0,625384	30,00%
B1 – Residencial Baixa Renda Normal			
Consumo até 30 kWh	0,110942	0,162106	30,00%
Consumo de 31 a 80 kWh	0,248617	0,363274	30,00%
Consumo de 81 a 100 kWh	0,249500	0,364566	30,00%
Consumo de 101 a 180 kWh	0,374235	0,546826	30,00%
Consumo acima de 180 kWh	0,415834	0,607610	30,00%
B2 – Rural Normal			
Independente do Consumo	0,250448	0,308969	18,00%

national inequities in economic status. One is the Reserva Geral de Garantia (the RGG, established in 1977) which created a uniform tariff by transferring money from the lower cost and more profitable companies (in the South) to the higher cost companies serving lower income populations in the North. Another fund, the Global Reserve for Reversion (RGR), financed by a 3% levy on fixed assets, is intended to fund new construction and has been spent primarily on rural electrification.

A number of government programs have been put in place to improve the electrification situation in Brazil. The Luz no Campo program was aimed at extending the national grid system into adjoining areas that were unelectrified. The Programa de Desenvolvimento Energético de Estados e Municipios (PRODEEM), discussed further below, specifically targeted community structures such as schools for electrification. Due to the remoteness of the communities for which PRODEEM was established, electrification was done through solar panels.

Until recently, however, the concessionaires did not make all the effort required to reach universal service, prompting legislative action and ANEEL to establish a 2015 deadline (ANEEL 2003). As part of the larger rural welfare programs of the Lula government, a law (Luz Para Todos – Light for All) was passed that provided financial incentives for utilities to achieve their universal access goals if they met more stringent targets (2008 instead of 2015).[1] (O Governo de Brasil 2003; Ministério de Minas e Energia 2004)

The explicit goal of the program is universal electrification with a fixed date to meet the program's objectives:

Art. 1o- Fica instituído o Programa Nacional de Universalização do Acesso e Uso da Energia Elétrica – "LUZ PARA TODOS," destinado a propiciar, até o ano de 2008, o atendimento em energia elétrica à parcela da população do meio rural brasileiro que ainda não possui acesso a esse serviço público (O Governo de Brasil 2003).

Art. 1 – A National Program for Universalization and Use of Electrical Energy – "Brazil:Luz Para Todos" shall be instituted, destined to provide by the year 2008, electricity service to the portion of the population in rural areas of Brasil that still do not possess access to this public service (Author's translation).

[1] Importantly, this is done at no cost to the consumer. The concessionaire can charge regulated tariffs, but cannot charge for connection under Luz Para Todos.

The program provides funding to the Brazilian utilities to meet their regulatory obligations to serve everyone within their exclusive service territory. Universal service in this case means primarily households. This reflects the broader emphasis internationally, where the most used statistic used is the proportion of the population that is unelectrified.

The possibility of receiving funds for expansion rather than the unfunded mandate of ANEEL led the utilities to sign on to the basic bargain (Br.Ind September 2005; Br.Acad.SB September 2005; Br.MME September 2005). They now have access to significant federal resources for connecting new customers (or building isolated or individual systems for households too far from the grid). Of the $2.4 billion dollars estimated by the Ministry of Mines and Energy necessary for universal electrification, the federal government will provide 72% of the funds.[2] This is primarily through the grants of the Conta de Desenvolvimento Energético (CDE) and RGR grants and loans. (Br.MME September 2005) While it is becoming increasingly clear that this target will only be met in the more industrialized south, where the challenges are not as great, the government remains committed to the goal (Br.Acad.AM September 2005; Br.Acad.SP September 2005; Br.ANEEL September 2005; Br.Donor September 2005; Br.MME September 2005). The Luz Para Todos funds alleviates one of the financial burdens utilities face in reaching rural customers, the high capital costs. The difference between the low tariffs for rural customers and the high cost of service continue to be covered by other mechanisms, primarily different forms of cross-subsidization. One of those additional sources of revenue is the Conta de Consumo de Combustiveis (CCC), a federal subsidy for the use of diesel fuel. The CCC diesel subsidy amounted to Rs 3.5 billion in 2005 with 80% going to five state capitals in the Amazon region. This further skews the economics towards centralized utilities (that can access the CCC) over local actors and diesel options over other technologies (Br.MME September 2005).

Table 3.2 and Table 3.3 provide a summary of the major actors, laws, regulations and programs relevant to rural electrification in Brazil. The next section describe the distributed rural electrification efforts in Brazil by two very different sorts of actors. First are the centralized utilities, which have used DG technologies to meet their universal service obligations. The second sub-section outlines the large variety of programs that have been put in place to fill the gaps left by the centralized system.

As a result of the institutional structure of the Brazilian electricity system, particularly its regulations mandating exclusive service territories, the distributed electrification efforts in Brazil can be divided into two groups for purpose of analysis. The first, and in many ways, dominant group are the highly centralized efforts of the utilities and the central government (often working through the utilities). Their dominance is the result not only of the regulatory system but also of the tariff structure and of the money flows from the government. They are centralized in terms of organization, but have variation in the other business model factors (e.g. technology, target customers and financing).

---

[2] At 3 R$/US$

Table 3.2 Major actors in Brazil's rural electrification

Organization	Function	Comments
Agência Nacional de Energia Elétrica (ANEEL)	Regulation of all aspects of the electricity industry	Enforces conditions of exclusive service territories, technical requirements, tariffs for low income consumers, etc.
Cooperatives	Cooperatives have been in place for a long time in Brazil and theoretically have the right to become concessionaires under ANEEL regulations	The technical requirements for becoming a concessionaire may cause some cooperatives to give up independent electricity supply
Eletrobras and subsidiaries	Government owned holding company and subsidiaries in generation, transmission and distribution	Own utilities in the less profitable and more challenging rural areas of the north of Brazil, including the Amazon
Government distributors	Distribution of electricity to a concession area	Have exclusive service territories and obligations to serve as set by ANEEL. Even though three of the six federal distributors made money in 2004, federal distribution resulted in net losses to Eletrobras
International Donors	Funding of rural electrification efforts	Have been responsible for early failures in diffusing technology. Heavily involved in promoting restructuring in Brazil
Ministério de Minas e Energia (MME)	Oversees national energy programs, including the Luz Para Todos program for universal service	
Non-governmental organizations	Have been involved in a number of rural electrification projects	Facing stiff competition from utilities meeting their concession obligations, either because they are not official concessionaires or because their stand-alone systems can't compete with the high subsidies of the centralized system
Private distributors	Distribution of electricity to a concession area	Have exclusive service territories and obligations to serve as set by ANEEL
Private enterprises	Have provided electricity in areas that the utilities have not served (e.g. diesel mini-grid entrepreneurs)	Illegal under ANEEL regulations and charge tariffs that don't meet ANEEL's requirements

The deficiencies in the centralized model, particularly the lack of incentives until recently to fully serve the consumers within their service territory, have led to the development of a number of alternative models. These models are not able to access many of the same resources as the centralized utilities (either in terms of government funding or in terms of a diverse consumer base that can be used for cross-subsidization) and so they remain on the periphery of the distributed electrification effort.

Table 3.3 Major legal documents, government programs and subsidies relevant to rural electrification

Program/subsidy	Function	Comments
ANEEL Resolution 223 (2003)	Fixes deadlines for concessionaires to meet their universal service obligations	Deadlines range from 2006 to 2015 depending on the current level of electrification
CCC	Conta de Consumo de Combustiveis. To provide subsidized diesel, particularly in the Amazon	Makes diesel competitive for those with access to the CCC. Not all consumers qualify
Decreto 4.873 (2003) and Portaria No. 447 (2004)	Established the Luz Para Todos universal electrification program that provides incentives for utilities to meet their obligations by 2008	Utilities have flexibility to meet obligations via both grid extension and centrally managed DG. Customers don't pay for installation and their tariffs are set by ANEEL
Lei 10.438 (2002)	Establishes Programa de Incentivo às Fontes Alternativas de Energia Elétrica (PROINFA) for promotion of renewables and the Conta de Desenvolvimento Energético (CDE) for funding universal electrification	It also sets the limits for who is eligible for reduced tariffs
Lei 10.762 (2003)	Makes technical modifications to earlier laws, including 10.438	
Low income tariffs	To reduce the financial burden on low income families of having electricity service. Tariffs are set according to consumption level per household with households below 30 kWh/month paying a highly reduced rate (35%), those between 30 and 100 kWh paying 60% and those between 100 kWh and a regional limit (~240 kWh/month) paying 90% of the residential tariff	The difference is made up through cross-subsidies (private utilities) or by passing losses onto the federal holding company (government utilities)
Luz No Campo	A grid-based universal electrification program	Did increase electrification rates but was not designed to meet the needs of the most remote populations
PRODEEM	Programa de Desenvolvimento Energético de Estados e Municipios. Promoted the use of renewables for community structures such as schools and clinics	Has had mixed success in being able to install and, more importantly, maintain systems
RGG	Reserva Geral de Garantia	Forced a uniformization of tariffs between more expensive northern utilities and more profitable southern utilities
RGR	Reserva Global de Reversão. A levy on fixed assets to finance continued construction in the electricity sector	Money collected spent primarily on rural electrification

However, there is much to be learned concerning the possibility of rural electricity markets and the contributions that electricity can play to larger development efforts. It should be noted that there is some movement on the institutional front. While the Luz Para Todos program has been focused exclusively on serving households, there has been some call to expand to include productive activities. This has required the energy ministry to look into partnering with NGOs, the Ministry of Rural Development and agriculture programs through PRONAF (Br.MME September 2005).

Centralized Organizations Delivering Distributed Power

Overview

The use of DG in Brazil for rural electrification is dominated by the installation of diesel mini-grids and solar home systems by centralized utilities. This centralized utility model is a direct result of the regulations governing the electricity sector which mandate exclusive service territories for concessionaires and low tariffs for lower income consumers. The policies of the federal government have reinforced the centralized utility model through the Luz Para Todos program.

Companies like the Companhia Energética do Amazonas or CEAM (the utility serving the state of Amazonas) are government owned utilities. CEAM serves the rural areas of Amazonia and is installing more mini-grids based on diesel generation to serve its rural customers. The capacity of the CEAM grids allow for basic household electrification, but not much else. As the official concessionaire, CEAM is able to utilize the Luz Para Todos funds plus the CCC diesel subsidy to keep costs down. However, since all of its customers are rural and lower income, CEAM also relies on its status as a government utility, part of the Eletrobras group, which allows it to run a deficit (Br.Site.Aq September 2005). CEAM's losses in 2004 (before taxes, interest, depreciation and amortization) were R$ 71 million (approximately $35 million U.S.) (Eletrobras 2005)

Others like the Companhia de Electricidade do Estado da Bahia (COELBA), in Bahia, are private utilities operating under a concession agreement with the federal government. Like other utilities, COELBA is relying primarily on extending their grid wherever possible (the lowest cost option for most of their customers) but using solar home systems to provide basic electricity services for their most remote and dispersed customers. These solar home systems are installed and maintained by CEAM and customers pay a reduced tariff similar to that paid by low-income grid-connected customers. The focus, again, is on household electrification. Like CEAM, they can utilize Luz Para Todos funds for construction. However, COELBA is a privately owned utility and does not have a higher level government holding company to absorb the losses. What it does have is a more diverse mix of customers and it is using richer customers to cross-subsidize its poorer customers, including those receiving solar home systems (Br Ind September 2005).

This is not to say that the utilities did not undertake any action prior to the Luz Para Todos program. The utility of Minas Gerais, the compartia Energética de Minas Gerais (CEMIG), undertook a rural electrification program that included the use of PV. The program had components for electrifying community structures, schools and households. For the household program, CEMIG covered nearly two third of the capital cost and the municipality within which the community was located would cover the other third. Households had to pay a monthly flat fee which was sufficient to cover eventual battery replacement, but not enough to cover all operations and maintenance, which CEMIG ended up paying for. As of 2001, CEMIG had fallen short of its original goals. Between 1995 and 2001, 450 solar home systems were installed of the estimated 4,700 expected. CEMIG also found itself paying more than its anticipated share of the capital costs due to poor municipal finances. The program was switched to a consumption based tariff on par with its grid customers (Diniz et al. 1998; Diniz et al. 2002).

The PRODEEM program established in 1994 used PV systems to electrify rural communal structures such as schools, clinics, etc. Systems were given to communities and little provision was made for operations and maintenance. While a large number of systems were installed over the six phases of the program between 1994 and 2002, PRODEEM was also plagued with a number of problems. Some were due to poor equipment, but many were also due to poorly chosen equipment (e.g. undersized inverters) or institutional problems. The result is that an unknown number of the PRODEEM installations are no longer working. One review sampled a small number of units (79 out of the 8,700 installed at the time) and found upwards of 50% not working (Galdino and Lima 2002; ESMAP 2005).

Utility Diesel Mini-Grids

Utilities serving remote communities, such as CEAM in Amazonas State, are often relying on diesel generators to provide electricity through mini-grids. These diesel mini-grids are installed, serviced, and operated by the central utility and service the entire community, including households, community structures and some productive activities (such as shops). However, as the focus is on household electrification, the systems run close to capacity and are not designed to handle major productive loads.

CEAM, like other utilities in the north, is making progress in electrifying communities within its service territory. Twenty-five new systems were to be installed in 2006. The units were expected to range in size from 80 kW to 20.6 MW with an average installation size of 2.24 MW. This can be compared to an average size of existing units of roughly 700 kW. CEAM, as a public utility owned by Eletronorte (itself owned by Eletrobras) can rely on soft budget constraints in addition to the loans and grants available through the Luz Para Todos program in order to meet its obligations for universal service. Operations and maintenance costs are covered by tariffs plus two subsidies. First, the fuel is subsidized through the CCC. Second, CEAM uses cross-subsidies, as mandated by the regulations. The cost of supply is Rs 150/MWh or Rs 0.15/kWh with the CCC and double without, higher than the rural tariff and significantly higher than the low-income residential tariff. That does not account for the transport up the river (which can take days by barge). As a result, CEAM is losing money, which it passes on to its holding company (effectively resulting in another cross-subsidy with richer consumers from the south that buy the electricity from the generation units owned by the holding companies).

While CEAM is installing new units, progress is slow and they are not expected to meet the 2008 deadlines set under the Luz Para Todos program. This is also true for other utilities serving the northern part of Brazil, where the largest proportion of unelectrified people live and where the unelectrified population tends to be more remote (Eletrobras 2005; GTON 2006; Br.ANEEL September 2005; Br.Site.Aq September 2005)

DREM Parameters

Organization: The diesel mini-grids are installed and operated by the centralized utility responsible for that region.
Technology: Diesel generator with mini-grid.
Target Customers: The entire village.
Financial Structure (Capital): Utilities can use funds from the Luz Para Todos program for capital costs. Capital costs can also be covered through soft budget constraints by utilities within the public holding companies.
Financial Structure (O&M): Below-cost tariffs are cross-subsidized by urban consumers, the soft budget constraint of the public utilities and the diesel subsidy.

Control Variables

Capital Cost Subsidies (HIGH): The utilities can access the subsidies provided through the Luz Para Todos program as well as through internal financial transfers. These subsidies cover a significant portion of the costs.
Operating Cost Subsidies (HIGH): The utilities can subsidize O&M through cross-subsidies or soft budgets. Tariffs are set very low by the regulations so the operating cost subsidies cover a significant portion of the cost of maintaining the systems.
Density of Customers (MEDIUM): Customers are clustered in relatively compact villages.
Remoteness of Customers (HIGH): Customers are remote (e.g. accessible only by boat).
Policy Regime (FAVORABLE): The policy regime is favorable as the federal government has both mandated service and provided financial incentives to the utilities.
Regulatory Regime (FAVORABLE): The regulatory regime is favorable as utilities have an obligation to serve all customers within their exclusive service territory by whatever set of technologies is appropriate. However, regulations for mini-grid systems have not yet been published. Grid based regulations would be difficult for mini-grid operators to meet and may be unnecessarily too strict (e.g. 24 h service).

Outcomes

Access (HIGH): Given the difficulties in serving the rural population through the grid system, diesel based mini-grids will remain a key component of the utilities portfolios.

Sufficiency (MEDIUM): The utility mini-grids are designed to meet basic levels of electrification for households, though they can include community structures. However, the incentive is not high to provide more power since the mini-grids operate at a loss. This can lead to the exclusion of productive activities.

Quality (HIGH): Utilities are required to meet quality regulations.

Sustainability (HIGH): The utilities are obligated to provide service under the existing regulations. So long as the regulations are in place and costs can be absorbed, these mini-grids will continue to function.

Replicability (HIGH): Replicability is high since mini-grids are a likely solution for a number of communities that the utilities must serve.

COELBA Utility Solar Home Systems

Like all Brazilian utilities, COELBA has to meet commitment under ANEEL regulations to electrify all of the customers within its service territory (the state of Bahia). The utility has 3.5 million customers, of which half are low income (including in urban centres). Using the incentives under the Luz Para Todos program, COELBA intends to electrify 360,000 consumers by 2008. Of this number, they expect to serve 10% by installing solar home systems. COELBA is paying 20% of the cost with the rest from Eletrobras and other Luz Para Todos funds. According to COELBA, it would be difficult to serve these customers without LPT and they would have to increase tariffs to cover the high costs (Br.Ind September 2005).

However, a study published in 2000 estimated a market for Bahia of 170,000 systems *without* a subsidy. The study noted the high potential for solar home systems in Brazil, given the remote populations and good solar insolation. The results showed the potential for a relatively large market, with a 30% subsidy resulting in almost 50% more households able to afford the systems. However, they also found that even with 100% subsidy of equipment, the poorest two income groups would not be able to afford the monthly fee (ESMAP 2000).

COELBA's rule of thumb is that when the cost of extending the grid line is >R$13,000 and the load is not too large, then they will install a solar system. They expected to install 3,000 SHS in 2005 and the rest in the 2006–2008 timeframe. The systems are sized to supply 13 kWh/month, enough for three lamps, a TV, and a radio. The consumers pay the normal tariff (R$ 2/month), which includes the low consumption discount. Without the subsidy, the price would be R$20/month. Maintenance of the solar home systems is done by COELBA. Trained employees are at the closest distribution substations (Br.Ind September 2005).

There has been some issues with sending bills. COELBA arranged with ANEEL that customers can go to the utility, request the bill and pay at the same time. Evidence from COELBA is that, despite their low incomes, the subsidized customers had high bill payment rates. This was attributed to being both a matter of pride and the practical reason that it gives them a legal document to show residence (Br. Ind September 2005).

DREM Parameters

Organization: A privatized utility serving the state of Bahia.
Technology: Solar home systems
Target Customers: Households
Financial Structure (Capital): Grants and loans through the Luz Para Todos program plus internal cross-subsidies.
Financial Structure (O&M): Tariffs plus cross-subsidies from urban consumers.

Control Variables

Capital Cost Subsidies (HIGH): The utilities can access the subsidies provided through the Luz Para Todos program as well as through internal financial transfers. These subsidies cover a significant portion of the costs.
Operating Cost Subsidies (HIGH): The utilities can subsidize O&M through cross-subsidies or soft budgets. Tariffs are set very low by the regulations so the operating cost subsidies cover a significant portion of the cost of maintaining the systems.
Density of Customers (LOW): Insufficient customers to justify grid
Remoteness of Customers (HIGH): Customers are remote.
Policy Regime (FAVORABLE): The policy regime is favorable as the federal government has both mandated service and provided financial incentives to the utilities.
Regulatory Regime (FAVORABLE): The regulatory regime is favorable as utilities have an obligation to serve all customers within their exclusive service territory by whatever set of technologies is appropriate.

Outcomes

Access (HIGH): Tens of thousands of households will be electrified through solar home systems under COELBA's plans to meet its Luz Para Todos obligations.
Sufficiency (LOW): Solar home systems are limited to providing basic lighting and entertainment services.
Quality (HIGH): Solar home systems, if maintained, can provide steady power.
Sustainability (HIGH): The utilities are obligated to provide service under the existing regulations. So long as the regulations are in place and costs can be absorbed, solar home systems will be part of the portfolio.

Replicability (HIGH): Replicability is high since solar home systems are a least cost solution for a number of households that the utilities must serve.

CEMIG Utility Solar Program

The utility in Minas Gerais (CEMIG) started a solar PV program prior to the establishment of Luz Para Todos. It consisted of three separate components. First was a community electrification program which included water pumping. It was targeted at high priority rural areas and was intended to cover households, community structures and water pumping. It was set up in partnership with CEPEL, NREL, and MME (through the PRODEEM). Prefectures were responsible for paying CEMIG trained maintenance person when there is problem. It was piloted in 1996 and then a second phase of 13 communities was rolled out. The plan (as of 1998) had been to expand into 450 communities in Minas Gerais.

Second was a separate program targeted at schools with plans for 100 schools between 1997 and 2000 in partnership with the state and federal governments. The goal was to provide both electricity and water to schools (with water also for the community in some cases).

The third leg was a household electrification program (Luz de Minas), established in 1995 with a goal of electrifying 100 k households. A small fraction (4,700) was to be supplied with solar home systems. There are households that are currently using kerosene and dry cells and paying between $4–10 per month. CEMIG would cover 64% and the municipality (not the community) would pay 36%. The users would pay a monthly fee ($4 or $8 for 50 W or 100 W systems respectively). This covers the cost of battery replacement but does not cover the cost of technicians.

CEMIG established criteria to ensure cost-effectiveness as compared to the grid and a training program for all stakeholders. However, despite these measures, the program fell short of its original goals. As of 2001 there were 115 community PV systems installed and 450 household systems installed. CEMIG also decided to switch to a consumption based tariff on par with the grid-connected consumers. Finally, CEMIG was unable to get the necessary funds from the municipalities due to their own budgetary limitations and so the utility anticipated paying up to 90% of the investment costs at the time (Diniz et al. 1998).

DREM Parameters

Organization: Centralized Utility
Technology: Solar systems
Target Customers: Households, community structures, and water pumping
Financial Structure (Capital): Provided by the utility and the municipality
Financial Structure (O&M): User fees to cover replacement parts but not technicians

Control Variables

Capital Cost Subsidies (HIGH): Customers were not responsible for any capital costs

Operating Cost Subsidies (MEDIUM): Customers covered some of the operating costs through a fee.

Density of Customers (MEDIUM): Communities were dense enough to support schools or other community structures requiring electrification, but it was clearly better for the program to provide stand-alone systems rather than micro-grids.

Remoteness of Customers (HIGH): CEMIG had to resort to these solar systems because the customers were not within reach of their planned grid extensions.

Policy Regime (FAVORABLE): The project was part of PRODEEM and supported by the government.

Regulatory Regime (FAVORABLE): The project was by the state utility which has a monopoly over its territory.

Outcomes

Access (LOW): The targets were not met and in the end only 450 household and 115 community systems were installed by 2001, roughly 10% of their stated goals.

Sustainability (LOW): The fact that CEMIG switched to a tariff based on grid prices shows that the original model was not sustainable.

Replicability (LOW): CEMIG was not able to get the municipalities to cover their share of capital costs and had to cover 90% of the costs rather than 64%. The number of installations did not meet the goals, indicating that CEMIG itself was not able to replicate the model within its territory, making the likelihood of repeating it elsewhere low.

PRODEEM

The Program for Energy Development of States and Municipalities (Programa de Desenvolvimento Energetico de Estados e Municipios, PRODEEM) was established by Presidential Decree in 1994. Its purpose was the use of decentralized electricity for community electrification. It was not intended to electrify households, but rather to provide benefits to communities by electrifying communal structures such as schools and clinics.

The technology chosen was photovoltaic panels and by 2000, the program provided electricity in 3,050 communities, expected another 1,050 in 2000, and 1,086 in 2001 (plus another 3,000 tendered). Through Phase V of the program a total of 8,742 systems were installed (totaling 5,209 kW). Of this, the majority (5,914 systems) were PV systems for electrification of communal buildings, about a quarter were for water pumping (2,449 systems) and a few were for public lighting (379 systems).

Units were provided free of charge to communities with no provision for cost-recovery for maintenance and minimal community involvement. A later review of a small sample (79 systems) showed only 44 working. Overall, significant technical problems had to be overcome as the project went from phase to phase. Some were equipment problems, but many involved poor technology choices for the usage (e.g. undersized inverters or hard to replace batteries). This involved changes in batteries and other equipment, installation methods, and changes in responsibilities of the organizations involved. However, the fundamental problem of a lack of ongoing revenue to handle operations and maintenance costs (or institutional system in place to provide such services) is what likely led to the failure of a significant number of systems in the small sample, a result that would likely be replicated in a larger sample (Galdino and Lima 2002; ESMAP 2005).

DREM Parameters

Organization: Central Government
Technology: Photovoltaic systems
Target Customers: Community structures
Financial Structure (Capital): Federal Government
Financial Structure (O&M): None

Control Variables

Capital Cost Subsidies (HIGH): All capital costs were paid for by the government.
Operating Cost Subsidies (LOW): There are no indications any subsidies were provided for operating the system.
Density of Customers (MEDIUM): Small communities were the target recipients.
Remoteness of Customers (HIGH): The target was remote rural communities.
Policy Regime (FAVORABLE): This program was developed by the Ministry.
Regulatory Regime (FAVORABLE): Being a government program and using single installations rather than a micro-grid, there were no regulatory barriers.

Outcomes

Access (HIGH): A large number of systems were installed.
Sufficiency (HIGH): Systems were sized to meet the required load.
Quality (Unknown):
Sustainability (LOW): The lack of revenue from either tariffs or an operating subsidy has led to failures.
Replicability (LOW): This model would require the addition of large operating subsidies or a tariff system to be replicated (Table 3.4).

Table 3.4 Summary table for Brazilian distributed rural electrification models (Sources cited in main text and (ESMAP 2000; Correia et al. 2002; Gaube 2002; Winrock International Brazil 2002; Goldemberg et al. 2004; ESMAP 2005; GTON 2006; Int.Don6 August 2005; Int.Don2 July 2005; Br.Acad.AM September 2005; Br.Acad.SP September 2005; Br.ANEEL September 2005; Br.Donor September 2005)

DREM parameters		Utility diesel	Utility SHS	CEMIG solar program	PRODEEM
	Organization	Centralized Utility	Centralized utility	Centralized utility	Central government
	Target customers	Villages	Households	Community structures, households, water pumping	Community structures
	Technology	Diesel mini-grid	Solar home system	Solar systems	PV
	Financial: capital	Grants/loans/soft budget	Grants/loans/equity	Grants	Government program
	Financial: O&M	Tariffs/cross-subsidy/soft budget	Tariffs/cross-subsidy	Tariffs/cross-subsidy	No O&M recovery
Control variables	Capital cost subsidies	High	High	High	High
	Operating cost subsidies	High	High	Medium	Low
	Customer density	Medium	Low	Medium	Medium
	Customer remoteness	High	High	High	High
	Policy regime	Favorable	Favorable	Favorable	Favorable
	Regulatory regime	Favorable	Favorable	Favorable	Favorable
Outcomes	Access	High	High	Low	High
	Sufficiency	Medium	Low		High
	Quality	High	High		
	Sustainability	High	High	Low	Low
	Replicability	High	High	Low	Low
Notes on institutional factors	Policy measures	Luz Para Todos providing significant funds	Luz Para Todos providing significant funds	PRODEEM provided funds	Replicable as long as gov. willing to continue to fund.
	Regulatory measures	Regulatory requirements forcing electrification	Regulatory requirements forcing electrification	CEMIG was the monopoly supplier for the state.	

(continued)

Table 3.4 (continued)

	Utility diesel	Utility SHS	CEMIG solar program	PRODEEM
Other	Subsidies allow for high sustainability and replicability Subsidies and soft-budget constraints for CEAM make it affordable	Subsidies allow for high sustainability and replicability Subsidies make it affordable		

Alternatives to the Centralized Model

Overview

The limited progress centralized utilities have made until recently in rural electrification has lead to a number of alternative distributed models. However, these alternative models have been limited in their impact on rural electrification in Brazil. They cannot compete directly with the centralized utilities both because of the legal mandate and because of the tariff structure and subsidy system to keep rural prices low. The recent expansion of the centralized system as the result of the Luz Para Todos program calls into question the role these alternative models can play in future rural electrification efforts. To a certain degree the Brazilian government has recognized that the focus of the centralized utilities on basic household electrification is limited and started to develop integrated action plans to meet more general economic development needs. These action plans would utilize more distributed actors rather than the centralized utilities.

Table 3.5 provides a summary of the relevant characteristics and the outcomes of some alternative distributed electrification models in Brazil. A number of distributed electrification efforts are not included in these tables such as the private diesel generators, cooperatives and pilot projects to use various biomass sources (such as acai) in gasification systems. However, to the extent possible, information from these other cases, primarily based on interviews, was brought to bear in drawing the broader conclusions regarding distributed electrification in Brazil.

In the absence of action by the centralized utilities until recently, a number of alternative models for using DG technology have arisen in Brazil's rural areas. One NGO, Instituto para o Desenvolvimento de Energias Alternativas e da Auto Sustentabilidade (IDEAAS), has established a fee for service model to provide solar home systems in conjunction with its sister organization, a for-profit company. Customers pay an installation fee and a flat monthly fee (varying depending on the capacity of the system) in return for service. A combination of loans and grants are used to obtain capital. A few hundred systems have been installed, but financial sustainability has not yet been achieved (a minimum of 4,000 units would have to be installed). The recent push by the centralized utilities is causing IDEAAS to consider moving its focus to the northeast of Brazil where more people remain unserved and projections are that the utilities will not meet their deadlines of 2008. They are also looking at ways to partner with the utilities. The model would appear to be both sustainable and replicable in the absence of utility competition, but the costs limit their customer base to richer rural households (Int.Don6 August 2005; Mugica Undated).

Another NGO, BRASUS, is focused on providing renewable energy technologies to rural productive consumers and establishing a sustainable market for such technologies through Regional Market Managers. Capital, primarily from international donors, is used to set up a revolving fund and loans, carefully screened for credit-worthiness, are provided to the rural producers. The focus on productive

Table 3.5 Summary table for Brazilian distributed rural electrification models (continued) (Sources cited in main text (ESMAP 2000; Correia et al. 2002; Gaube 2002; Winrock International Brazil 2002; Goldemberg et al. 2004; ESMAP 2005; GTON 2006; Int.Don6 August 2005; Br.Acad.AM September 2005; Br.Acad.SP September 2005; Br.ANEEL September 2005; Br.Donor September 2005))

		BRASUS	IDEAAS SHS	SBC
DREM parameters	Organization	NGO plus regional coalition	NGO – for profit partnership	Entrepreneur plus NGO
	Target customers	Productive activities plus others	Richer households	Households
	Technology	Varies	Solar home system	Solar battery charging station
	Financial: capital	Loans	Loans/grants – installation fee	Grants
	Financial: O&M		Monthly fee	Fees
Control variables	Capital cost subsidies	Low	Low	High
	Operating cost subsidies	None	Low	None
	Density of customers	N/A	Low	Medium
	Remoteness of customers	High	High	High
	Policy regime	Neutral	Neutral	Neutral
	Regulatory regime	Unfavorable	Unfavorable	Unfavorable
Outcomes	Access	Low	Low	Low
	Sufficiency	High	Medium	Low
	Quality	High	High	Low
	Sustainability	High	High	Low
	Replicability	High	Medium	Low
Notes on institutional factors	Policy measures	Integrated Action plans of MME envision partnering with NGOs on productive activities	LPT reducing incentive for individuals to obtain SHS since connection is free under LPT	
	Regulatory measures		Universalization requirements on utilities bringing them into competition with IDEAAS	
	Other			Frequent recharging Expensive

activities (particularly agricultural processing) increases BRASUS' sustainability and also makes it less susceptible to competition from the utilities' expansion plans since many of those are focused on serving households (BRASUS 2005a, b).

Another entrepreneur based model was based on solar battery charging stations. This project by an international donor used local entrepreneurs to run the stations. However, this model failed as customer dissatisfaction with the service rose over time. Customers complained about the need to bring their batteries to the station, the fact that battery life decreased quickly (necessitating more frequent visits and a rise in their monthly expenses), and accused the entrepreneurs at times of favoritism in handing out batteries.[3] (Santos and Zilles 2001)

Models based on local entrepreneurs have also been attempted. Within CEAM's service territory in the Amazon, numerous local entrepreneurs have installed small diesel generators to serve their and their neighbors basic electricity needs. Unfortunately, little data exists on these installations as they are outside the formal regulatory and legal system. However, a survey of 100 communities done by MME for CEAM found that 95 had a diesel generator (Br.MME September 2005). Indications are that costs of electricity are high, service is only in the evening and quality is likely low. One system in Nossa Senhora de Gracas in Amazonia had monthly charges that would be equivalent to 25–50 c/kWh (depending on usage) (Br.Site.NSG September 2005). It would appear likely that many of the elements of this DREM are similar to the rural electricity entrepreneurs operating diesel mini-grids in Cambodia, for which there is much more information.

IDEAAS/STA Solar Home System

The IDEAAS model is a rental model for solar home systems. The model is implemented by an NGO (IDEAAS) and a for-profit company called STA. However, IDEAAS and STA are, in fact, related organizations and share personnel. STA provides other electricity services (namely alarm systems) that account for the bulk of its profits. IDEAAS owns and sells SHS services with STA contracted to provide the maintenance.

Customers pay an installation fee plus a monthly fee and both vary according to the size of the system installed. Three systems are offered: 60 W ($180 plus $10/month); 90 W ($220 plus $16/month) and 120 W ($250 plus $21/month). One year of financing is available for the installation fee at roughly 3–5% interest. Since each service call costs ~$30 or about 3 months of maintenance fees, the groups are working to minimize service calls through training of the households. The households that are targeted are already spending a minimum of $11/month on energy and are at the

---

[3] Unlike other battery charging schemes, in this one the customer did not own a specific battery, but would come and exchange their battery for another one, thereby eliminating the need to come back and pick up their battery. However, different users had different usage patterns and over time battery performance began to vary widely.

higher end of the rural income scale. In June 2004 the first 40 units were installed. Another 200 units were then purchased from grant money. The projection is that the operation would be profitable after 4,000–6,000 units are installed (sources differ). Estimates were that the project would do so after 4 years operation.

The model does face some challenges, including the increased emphasis on universal electrification by the utilities as a result of the Luz Para Todos and related government laws and regulations. The initial focus of operations was in the southern part of the country where the utilities are more likely to meet their universal goals. One response has been to consider shifting services to the northern regions where utilities lag in meeting their commitments. There is also the possibility of partnering with a utility, however, this raises the problem of meeting the technical standards that the utilities are forced to abide by in the regulations (Bornstein 2007; Sutton 2007; Int.Don6 August 2005; Mugica Undated).

DREM Parameters

Organization: Non-governmental organization partnered with a for-profit enterprise.
Technology: Solar home systems.
Target Customers: Richer households
Financial Structure (Capital): Loans and grants plus installation fees
Financial Structure (O&M): Monthly fee

Control Variables

Capital Cost Subsidies (LOW): There is some support for the NGO from outside plus there is a profit-making wing that appears to provide limited cross-subsidization.
Operating Cost Subsidies (LOW): There is some support for the NGO from outside plus there is a profit-making wing that appears to provide limited cross-subsidization.
Density of Customers (LOW): The target customers are single households.
Remoteness of Customers (HIGH): The target customers are in the remote rural areas of the country.
Policy Regime (NEUTRAL): There is nothing that favors or disfavors this model within the policies.
Regulatory Regime (UNFAVORABLE): It will be hard to compete with utilities providing nearly free electricity to households as part of their regulatory mandate.

Outcomes

Access (LOW): The number of systems installed is low compared to centralized programs.
Sufficiency (MEDIUM): Multiple system sizes were offered, allowing people to chose the system that meets their needs. However, even the largest system is limited in the electricity services it could supply.

Quality (HIGH): A system is in place to ensure ongoing service.

Sustainability (HIGH): There is nothing to indicate that households currently using the solar home systems will give them up, even if the utility moves in.

Replicability (MEDIUM): While current users may continue, replicability may be made difficult in regions where the utilities are working to meet their obligations. However, since the deadlines for electrification are not realistic for some parts of the country, IDEAAS should still be able to expand for a while.

BRASUS

BRASUS is a Brazilian non-governmental organization that has set up a loan mechanism for renewable energy technologies. The organization targets productive activities, consumers that have the financial means to repay the loans. The initial funds for the revolving investment fund came from international donors, with the state of Mato Grosso also being involved. The goal is to create sustainable renewable energy markets. In order to do this, the NGO has created Regional Market Managers, which are a "consortium of private local organizations with complementary strengths responsible for working together and with other public and private partners they specifically involve to promote sustainable RE [renewable energy] markets in their region that will benefit their own (complementary) goals." RMMs provides institutional support (e.g. training, financial mechanisms, planning tools) to those wanting to install renewable energy.

BRASUS looks for win-win situations and so funds projects that enhance economic activities through installation of renewable energy technologies. However, these customers are combined with residential and community loads to enhance viability. Loans are at 6% or 8.75% (depending on size of lendee) with R$30,000 upper limit. Loan approval requires an evaluation of viability and collateral and loans are to be repaid in 5 years. Interest from loans will be used to fund the administration of loan contracting organization established by BRASUS. As of 2005, 47 projects had been funded and 19 organizations were participating as RMM members. The model had also been replicated from Mato Grosso to Piaui and 400 projects had been identified for short term (2005–2006) development.

Source: (BRASUS 2005a; BRASUS 2005b; Int.NGO October 2005)

DREM Parameters

Organization: Non-governmental organization with regional partnerships

Technology: Various

Target Customers: Primarily productive activities

Financial Structure (Capital): Loans are provided to the customers for purchasing equipment.

Financial Structure (O&M): Ongoing costs are the responsibility of the customer.

Control Variables

Capital Cost Subsidies (LOW): There is some international NGO involvement.
Operating Cost Subsidies (NONE): Some programmatic activities undertaken by international partners are covered by funding outside of the core business model.
Density of Customers: (N/A)
Remoteness of Customers (HIGH): These projects are undertaken in rural areas
Policy Regime (NEUTRAL): The policies in place neither support nor discourage these types of models. There is a growing recognition of the need to move beyond household electrification, so a more positive policy regime may be developing.
Regulatory Regime (UNFAVORABLE): The inclusion of neighboring commercial or household customers could violate the monopoly of the local utility company.

Outcomes

Access (LOW): Households are not targeted and so far only a limited number of productive actors have been provided with power.
Sufficiency (HIGH): Enough power is provided for the customers productive activities.
Sustainability (HIGH): The target customers are productive actors that have the income to support long-term maintenance and operations
Replicability (HIGH): Since BRASUS is targeted at productive activities, a group that the current utilities don't serve very well and are not obligated to serve in the same way as households, it is expected that the model should be replicable.

Solar Battery Charging Stations

Plans for solar battery charging stations in Brazil were initially introduced in 1996 by the Golden Genesis company. The plan was for 32 SBCs, 20 in Algoas, 10 in Bahia and 2 in Paraná. The stations would be 1 or 1.5 kW arrays, charging up to 12 batteries at a time and 50 or 70 home battery kits would be supplied per station. In order to avoid multiple trips to the station for drop off and pick up, a battery replacement service was offered, allowing customers to drop off one battery and pick up another at the same time.

Golden teamed up with a non-governmental organization and installed stations in Algoas and Bahia, while the Companhia Paranaense de Energia (COPEL) installed in Paraná. The stations were run by entrepreneurs that obtained a franchise. However, this model ran into difficulties. First, the tariff for battery charging ($2/charge) was insufficient to cover replacement parts. In addition, users complained about transporting the batteries and rapid reduction in battery lifetimes due to deep-discharge cycles. As a result, the number of battery charges per month went up (from 8 to 20), raising the cost to end-users. Finally, complaints about differences in batteries being handed back by entrepreneurs led to charges of favoritism.

In the end, all of the stations failed to operate profitably and most were shut down. Only in two cases in Algoas are a handful of customers still getting their batteries charged. Others resorted to alternatives (candles, kerosene, diesel, etc.) while some turned the battery charging into solar home system projects. In Algoas, the entrepreneur received a loan to buy systems and rent them to end-users. In Paraná, COPEL bought the systems and charges 25% to end user plus a monthly fee of $2.30. A calculation of levelized costs of SBC to SHS based on one year lifetime for SBC batteries and a 14% discount rate shows SHS are much cheaper ($3/kWh as compared to 4.41/kWh) (Santos and Zilles 2001).

DREM Parameters

Organization: Entrepreneur and non-governmental organization
Technology: Solar power battery charging
Target Customers: Households
Financial Structure (Capital): Provided by an international company and NGO.
Financial Structure (O&M): Fees for battery charging

Control Variables

Capital Cost Subsidies (LOW): Customers were not responsible for capital costs.
Operating Cost Subsidies (NONE): No operating cost subsidy was in place (despite operating costs being higher than the tariff charged).
Density of Customers (MEDIUM): Customers had to be close enough to bring their battery in for charging.
Remoteness of Customers (HIGH):
Policy Regime (NEUTRAL): There were no policies in place to support the project, but also no policies (other than the regulations) to stop them.
Regulatory Regime (UNFAVORABLE): The project could be put in place because the utilities were not meeting their obligations. However, they would appear to have been operating outside the regulatory rules.

Outcomes

Access (LOW): Total number of households that would have been served is less than 2000, a small number compared to those served by the utility solar programs (as just one counter-example).

Sufficiency (LOW): Batteries were only able to provide minimal levels of electrification.

Quality (LOW): There were ongoing complaints about battery quality

Sustainability (LOW): The project was not able to sustain itself. The few that remained fundamentally changed the business model in order to survive.

Replicability (LOW): The project was not replicated and the problems with battery replacement and with financing would have to be solved before it could be replicated.

Other DREMS

A number of other DREMs have been referenced in the literature, but there is insufficient information on the variables of interest. They are briefly summarized here:

- Cooperatives: There are a number of cooperatives that are involved in electricity generation and delivery from the pre-concession era. While some do have generation, most purchase electricity from concessionaire. At this point concessionaires can also now serve those areas, creating some conflict between the two types of entities. The result is a process of regularization of the cooperatives such that they become permissionaires under the current regulations. However, some are reluctant given the complexities of formalizing their systems and coming under the authority of the regulator. One example of a cooperative is APAEB (Associação de Desenvolvimento Sustentável e Solidário da Região Sisaleira), a cooperative of sisal producers that setup a revolving fund for solar home systems accessible by its members (Br.Acad.SB September 2005; Br.MME September 2005).

- There is a nascent solar home system market, estimated at 3 MWp per year by the Brazilian Renewable Energy Association (ABEER) (ESMAP 2005). For example, Kyocera is selling solar systems at R\$4,700/kW installed. However, now consumers can demand service from the utility so they are losing clients (Br.Ind September 2005).

- The NGO *Instituto Eco-Engenho* (IEE) supports solar home system entrepreneurs with cooperation of a development bank. They finance SHS entrepreneurs who charge a monthly fee for SHS (leasing model). There have been upwards of 90 entrepreneurs and 2,700 systems leased. However, defaults on payments have been high.

- PRODEEM estimates approximately 100 programs and projects. But most are pilot or university based projects.

- There are apparently a number of non-utility diesel mini-grids that have been established by local entrepreneurs or other local actors. One estimate is that 600 systems exist that do not get access to the CCC, but that includes some state concessionaire and municipally run systems in addition to those that are run by private entrepreneurs. As with private diesel systems elsewhere, the tariffs are quite high (up to $200/MWh) (ESMAP 2005). A survey of communities in the Amazon by the Ministry of Mines and Energy found that 95 out of the 100 unelectrified communities surveyed had a diesel mini-grid (Br.MME September 2005).

- Prior to the current COELBA solar home system effort, there was a donor program to install >10,000 SHS for which COELBA was contracted for the installation. Estimates are that 3,000 are still working. One problem was that consumers did not pay tariff (Br.Ind September 2005).
- The World Bank Rural Poverty Alleviation program ran from 2001 to 2005 and resulted in the installation of 16,000 solar home system units in 306 communities. A follow up survey found relatively high satisfaction (~65%) (World Bank 2005).
- The water company of Minas Gerais (COPASA) is providing water to remote communities through PV water pumps. Of the 7,200 communities needing pumping at the time the program was established, roughly 1,700 were communities greater than 2 km from CEMIG's grid. The water company set up a PV program with 48 systems installed by Dec. 1997. Operation and maintenance costs are controlled by the municipality (Diniz et al. 1998).
- CEAM has installed 170 SHS in 27 communities, each of 150 W. They are also looking to install other alternative energy sources. However, they have had mixed success, as evidenced by a biomass gasification system installed in Nossa Senhora de Gracas which had worked for about one hour in the eight months since installation (Br.Site.NSG September 2005).

Key Findings and Conclusions

As of 2009, 4 million Brazilians still did not have access to electricity. The use of distributed generation in Brazil for rural electrification is dominated by the installation by centralized utilities of diesel mini-grids and solar home systems. This centralized utility model is a direct result of the regulations governing the electricity sector and policies aimed at universal electrification. ANEEL, the Brazilian electricity regulator, oversees a system of monopoly service territories granted on a concession basis, much like in the United States. Concessionaires have rights (exclusivity in their service territory) and obligations (universal service). In addition, the tariff structure requires concessionaires to cross-subsidize their low-consuming customers (on the assumption that low consumption is correlated with low income) with higher tariffs for other consumers. Until recently, however, the concessionaires did not make all the effort required to reach universal service, prompting legislative action and ANEEL to establish a 2015 deadline. In absence of centralized electrification effort, small scale efforts arose to fill the holes in a potentially sustainable and replicable way. These range from entrepreneurs (e.g. small non-utility diesel grids in the Amazon) to coops and NGO efforts. The new laws and regulations are potentially putting the expanding utility system in conflict with these decentralized efforts.

The policies of the federal government have reinforced the centralized utility model. As part of the larger rural welfare programs of the Lula government, a law was passed that provided financial incentives for utilities to achieve their universal access goals if they met more stringent targets (2008 instead of 2015).[4] The possibility of receiving funds for expansion rather than the unfunded mandate of ANEEL led the utilities to sign on to the basic bargain. They now have access to significant federal resources for connecting new customers (or building isolated or individual systems for households too far from the grid). Of the $2.4 billion dollars estimated by the Ministry of Mines and Energy necessary for universal electrification, the federal government will provide 72%, through CDE grants and RGR grants and loans.[5]

The Brazilian experience with distributed rural electrification shows the power of centralized action coupled with relatively high subsidies. Access rates can be increased at a very affordable cost to even the poorest end consumer. However, while this model is sustainable and replicable within Brazil at the moment, it does require continued central government support through the full phase of the expansion. It also requires a viable tariff structure that allows cross-subsidization with full cost recovery and at tariffs for the subsidizers that does not cause widespread exit from the system. One does not have to look hard for examples in which this type of cross-subsidizing system suffers from both problems (e.g. many Indian utilities).

This combination of high costs and low revenues without either incentives or consequences led most utilities to move slowly to meet their universal service obligations. As with the other countries included in the larger study, the absence of strong central support to force widespread electrification left open a gap for alternative electrification models (e.g. private diesel operators, cooperatives, NGO providing alternative energies) to meet the needs of different consumers. In the absence of financial support from the central government, successful models have had to meet requirements for sound financial sustainability in other ways.

In contrast to the centralized utility model, these alternative models sometimes go beyond basic electrification. Some are focused on providing electricity to productive activities in order to improve economic output and development (e.g. BRASUS). Others remain focused on households (e.g. IDEAAS), but allow for higher levels of electricity consumption than the basic levels provided by the utilities and are decentralized in both technology and organization. The rural diesel

---

[4] Importantly, this is done at no cost to the consumer. The concessionaire can charge regulated tariffs, but cannot charge for connection under Luz Para Todos.

[5] At 3 R$/US$

entrepreneurs appear to operate much like those in Cambodia, selling power at high prices but to consumers who utilize it for the most basic of services and for a few hours a day, thereby reducing their monthly expenses.

New incentives are expanding the utility's geographic reach. However, the focus is on basic household electrification and tariffs have been kept artificially low, necessitating substantial ongoing subsidies. The incentives and the hard deadlines of the Luz Para Todos (even if they are delayed) will create problems for both existing and new models that may have advantages over the long-term and serve needs not being met by the centralized utilities.

The importance of institutional factors is also readily evident in the Brazilian case studies. The Luz Para Todos program, tariff structure, diesel subsidy to utilities and regulated concession areas create an environment in which utility based models are both favored to install systems and are able to sustain themselves despite the poor economics of rural electrification. The example of the failed solar battery charging stations also shows how institutional factors can interfere with a potentially viable business model. In this case, a combination of technical factors (the shortened battery life due to battery changing) and certain normative attitudes regarding the ease by which electricity should be delivered made this model ultimately unsustainable.

Finally, it is important to note that the both renewable and non-renewable energy sources have been used with success in Brazil but also that some of the renewable energy models have met with failure. That success and failure, however, appears to have less to do with the technologies themselves than with the business models. More importantly, perhaps, is that the renewable energy technologies were often chosen for their ability to operate in remote areas and their relative cost advantages over grid extension rather than for explicit climate reasons.

References

ANEEL (2003) Resolução No. 223, de 29 de Abril de 2003, Agência Nacional de Energia Elétrica, Brasilia, Brasil

Bornstein D (2007) How to change the world: social entrepreneurs and the power of new ideas. Oxford University Press, New York

Br.Acad.AM (September 2005) Author interview with academic. Manaus, Amazonas, Brasil

Br.Acad.SB (September 2005) Author interview with academic. Salvador, Bahia, Brasil

Br.Acad.SP (September 2005) Author interview with academic. Manaus, Amazonas, Brasil

Br.ANEEL (September 2005) Author interview with Agência Nacional de Energia Elétrica (ANEEL). Brasilia, Brasil

Br.Donor (September 2005) Author interview with bilateral foreign donor. Salvador, Bahia, Brasil

Br.Ind (September 2005) Author interview with COELBA – electric utility of the State of Bahia. Salvador, Bahia, Brasil

Br.MME (September 2005) Author interview with Ministry of Mines and Energy (MME). Brasilia, Brasil

Br.Site.Aq (September 2005) Author site visit to Aquidabam. Aquidabam, Amazonas, Brasil

Br.Site.NSG (September 2005) Author site visit to Nossa Senhora de Gracas. Nossa Senhora de Gracas, Amazonas, Brasil

BRASUS (2005a) Promoting sustainable markets for renewable energy in rural regions of Brazil. Solar World Congress, Orlando, FL

BRASUS (2005b) Regional market managers: a model for success in sustainable development, Brasil Sustentavel (BRASUS), p 9

Correia J, Valente A et al (eds.) (2002) A Universalização do Serviço de Energia Elétrica: Aspectos Jurídicos, Tecnológicos e Socioeconómicos. Salvador, Unifacs

O Governo de Brasil (2003) Atos do Poder Executivo, Decreto No 4.873, de 11 de Novembro de 2003, Diário Oficial da União, Brasilia, Brasil

de Oliveira A (2003) The political economy of the Brazilian power industry reform. Program on Energy and Sustainable Development, Stanford, CA, p 58

Diniz ASAC, Mendonça MSCC et al (1998) Current status and prospects of the photovoltaic rural electrification programmes in the State of Minas Gerais, Brazil. Prog Photovoltaics Res Appl 6:365–377

Diniz ASAC, França ED et al (2002) An utility's photovoltaic commercialization initiative: progress of the Luz Solar Programme for Rural Electrification. Twenty-Ninth IEEE Photovoltaic Specialists Conference

Eletrobras (2005) Annual report 2004. Rio de Janeiro, Eletrobras

Ministério de Minas e Energia (2004) Portaria No. 447, de 31 de Dezembro de 2004, Diário Oficial da União, Brasilia, Brasil

ESMAP (2000) Brazil rural electrification with renewable energy systems in the northeast: a preinvestment study. Joint UNDP/World Bank Energy Sector Management Assistance Programme (ESMAP), Washington, DC, p 114

ESMAP (2005) Brazil background study for a national rural electrification strategy: aiming for universal access. Energy Sector Management Assistance Program (ESMAP), Washington, DC, p 176

Galdino MA and Lima JHG (2002) PRODEEM – The Brazilian programme for rural electrification using photovoltaics. RIO 02 World Climate and Energy

Gaube J (2002) Producing electricity from renewable energy sources: energy sector framework in 15 countries in Asia, Africa and Latin America. Eschborn, Environmental Management, Water, Energy, Transport Division, Deutsche Gesellschaft für Technische Zusammenarbeit (GTZ) GmbH, p 191

Goldemberg J, Rovere LL et al (2004) Expanding access to electricity in Brazil. Energy Sustain Dev VIII(4):86–94

GTON (2006) Plano de Operação 2006 Sistemas Isolados. Rio de Janeiro, Grupo Técnico Operacional Da Região Norte, Centrais Eléctricas Brasileiras S.A. (Eletrobras), p 86

Hart SL (2005) Capitalism at the crossroads: the unlimited business opportunities in solving the world's most difficult problems. Wharton School Publishing, Upper Saddle River, NJ

IEA (2009) The Electricity Access Database. International Energy Agency. http://www.iea.org/weo/database-electricity/electricity-access-database.html

Int.Don2 (July 2005) Author interview with international donor. Washington, DC

Int.Don6 (August 2005) Author interview with multilateral donor. Washington, DC

Int.NGO (October 2005) Author interview with international NGO. Washington, DC

Mugica Y (undated) Distributed solar energy in Brazil: Fabio Rosa's approach to social entrepreneurship. UNC Kenan-Flagler Business School Cases, University of North Carolina Kenan-Flagler Business School, p 27

Paz LRLd, Fidelis da Silva N et al (2007) The paradigm of sustainability in the Brazilian energy sector. Renewable Sustainable Energy Rev 11(7):1558–1570

Santos RRd, Zilles R (2001) Photovoltaic residential electrification: a case study on solar battery charging stations in Brazil. Prog Photovoltaics Res Appl 9:445–453

Sutton CN (2007) The role of the utilities sector in expanding economic opportunity. Corporate Social Responsibility Initiative Report No. 24. Kennedy School of Government, Harvard University, Cambridge, MA

References

Winrock International Brazil (2002) Trade guide on renewable energy in Brazil. Winrock International Brazil sponsored by U.S. Agency for International Development, Salvador do Bahia, p 91

World Bank (2005) Implementation completion report on a loan in the amount of US$54.35 million to the State of Bahia for a rural poverty reduction project. International Bank for Reconstruction and Development, Washington, DC, p 56

Chapter 4
Distributed Rural Electrification in Cambodia

Keywords Cambodia • Diesel • Electricity entrepreneurs • Regulatory change

Introduction

Official estimates in 2004 were that 15% of Cambodians had electricity access. This would not be surprising given that the population is still largely rural or lives in smaller towns. Eighty percent of the population lives below the district town level and 75% live at the commune (or village) level (Cam.NGO.Dom March 2006). Électricité du Cambodge is responsible for roughly 9% of the electrified population and the other 6% is from licensed generators and distributors, primarily at the district level. This low level of electrification by the official utility and licensed entities is a legacy of the poor development and long-term military conflicts that have plagued this country. However, at the same time, official surveys show nearly 50% of Cambodian households owning a television set (National Institute of Statistics undated). More recent estimates are that 24% of households have lighting provided by either publicly or privately generated electricity, another 38% use batteries and 36% use kerosene (National Institute of Statistics 2009). However, another estimate is that 90–95% of the population may have access to a basic level of electricity, enough for a lightbulb in the evening (Cam.NGO.Dom March 2006). How then to explain this difference? The rest of the population is being served primarily by battery charging services and village level networks developed and run by small entrepreneurs. These Rural Electricity Entrepreneurs (REEs), as they are known, may supply 75–85% of the population and are completely unlicensed, unregulated, and are not supported by the government or donors (making exact numbers difficult to determine). This pattern of electrification in Cambodia presents an excellent opportunity to study a distributed electrification model that is widespread but usually at the margins of electricity supply in most countries.

Information on Distributed Rural Electrification Models in Cambodia comes from a combination of secondary sources and primary interviews conducted in Cambodia (December 2005 and March 2006) and Washington, DC (2005 and 2006).

While the REEs dominate the electricity supply situation in rural Cambodia, this is not the only model in use. Since the end of the military conflict, new models for rural electricity supply have been developed and implemented. As with many places, donors have established rural electricity projects. A cooperative has been established to use biomass for electricity supply at the village level. The commercial viability of the rural electricity market, as evidenced by the REEs, is also being tapped by a solar home system company which targets somewhat wealthier rural residents as well as non-governmental organizations operating in remote areas.

The Cambodian case, and the six Distributed Rural Electrification Models studied, is an important case because of the radical change in the institutional environment that occurred after the end of military conflict. During the decades of war and Khmer Rouge government, little was done centrally to develop the electricity system, especially in rural areas. Furthermore, there was no outside intervention. Private companies, non-governmental organizations and the large donors were not active in Cambodia during this time. This vacuum that existed during the war years (and for a period of time afterwards), was filled by local actions. In particular, local entrepreneurs arose to fill the void and provide a service they could see was necessary and potentially profitable. Their technology of choice was diesel generators (either with battery charging or with a combination of battery charging and mini-grids). There is significant evidence from other countries that this is an important component of the rural electrification landscape. However, it is one that is not often studied. The extensive use of the business model in Cambodia is an opportunity to learn more about this model and the conditions under which it operates.

The Institutional Environment

Like much else in Cambodia, the institutional environment for rural electrification and the pattern of electricity sector development have been heavily influenced by the years of military conflict in the country. The history of the sector can therefore be divided into pre-war, wartime and post-war eras, each with a different set of actors, institutional drivers and outcomes.

The electricity sector in the pre-war era was dominated, like in many countries, by a single utility whose initial focus was on the major urban centres. The electrification of Cambodia began a century ago with the formation of three electricity companies, two of which were purchased by the Cambodian government in 1958 to form Électricité du Cambodge (EDC). This government owned utility is still in existence and continues to be the single most dominant utility in Cambodia with responsibility for electricity transmission, importation of power from other countries, larger generating units and distribution to the larger provincial towns and the capital, Phnom Penh. Overall, however, it only serves a little less than 10% of the Cambodian population.

Development of the utility system was essentially halted during the many years of war. The unstable institutional environment caused by the decades of conflict meant that there were essentially no rules, no investment, and no planning in the electricity sector by the government. However, this low level of official electricity system development during the war years and shortly thereafter left open a gap that was filled primarily by small entrepreneurs.

The post-war era in Cambodia has seen some significant changes in the institutional environment for rural electrification. The grid system is being restored and expanded with the aid of foreign lending institutions and EDC is engaging in a number of projects to build infrastructure and import power.(World Bank 2003) In 2001, through the promulgation of a new Electricity Law, the government of Cambodia established an independent regulatory body, the Electricity Authority of Cambodia (Kingdom of Cambodia 2001; Int.Don4 January 2006). The EAC is responsible for licensing generation, transmission and distribution. What is most important from a rural supply perspective is the licenses for generation and distribution. The vast majority of operators in Cambodia are still unlicenced. However, the larger operators and those closest to the major cities are starting to obtain licenses. These licenses are often probationary and based upon expected upgrades of the distribution system to reduce losses, improve safety, etc. Even when they are not probationary, the license terms are for a short period of time (often five years) (Cam.Gov1 December 2005; Cam. NGO.Dom March 2006). However, importantly, the licenses do allow for tariffs that vary according to the needs of the private sector to recoup costs (a policy of the government of Cambodia) (MIME 2003; Int.Don4 January 2006).

One major driver for obtaining a license is the expansion plans of the major utility, Électricité du Cambodge.(World Bank 2003) EDC has developed a Master Plan which calls for expansion of the transmission system, development of new power plants and importation of power from neighboring countries. EDC plans to expand from the capital and provincial towns into the district towns and into communities along the planned transmission corridor. Without licenses, electricity entrepreneurs serving those areas would find themselves in the situation of operating illegally and in EDC's territory. By obtaining licenses, the hope seems to be that EDC will become primarily a wholesale seller of power in those communities and the pre-existing entrepreneurs will be allowed to continue as distributors (rather than generator-distributors) (World Bank 2003; Cam.NGO.Dom December 2005; Cam.Ind.REE3 March 2006; Cam.NGO.Dom March 2006).

Another characteristic of the post-war era has been the arrival of international donor aid, including both bilateral aid projects and aid from the multilateral institutions (e.g. the World Bank). One of the bilateral projects, donated by the Japanese government is discussed below as one of the business models existing in Cambodia. The World Bank has established a Rural Electrification Fund to help support various efforts to improve the access of rural populations to electricity (World Bank 2003; Int. Don4 January 2006). This will include decentralized efforts outside the EDC utility system. However, nearly 2 years after initial establishment of the fund, it had still not been operationalized, in part due to an inability to find a fund manager but also a desire to ensure safeguards were in place to prevent corruption (Int.Don3 April 2006; Cam.Gov2 December 2005; Int.Don4 January 2006). There has also been a lot of international interest in promoting various renewable energy technologies, as witnessed by various NGO projects and meetings dedicated to rural renewables. (Cam.Cons December 2005)

Cambodia thus presents a very interesting case. The evolution of the institutional structures that would ordinarily have developed to provide electricity service in rural areas was severely disrupted. Yet, in the absence of that centralized effort, individual actors moved to fill the need of the rural population. As a result, despite the poor economic conditions of rural Cambodia households, basic electrification is widespread. Of course, this is only possible because the households are willing to pay a high price for that basic service, despite their low incomes.

Rural Electricity Entrepreneurs

There are an estimated 600 mini-grids (these operators also usually run a battery charging service in addition to the grid) and 1,000 or more battery charging stations. While they are often considered separately, for the purposes of this analysis they are both grouped under the category of Rural Electricity Entrepreneurs (REE). Despite their differences in how they supply electricity to end-users, much of the technology (diesel generators) and the cost-recovering private sector nature of their business model makes them more similar than they are different.

A small sample survey of 45 REEs found they had on average 330 customers and start-up costs of around $21,000 (Enterprise Development Cambodia 2001). These Rural Electricity Entrepreneurs generally make a small profit, but to do so they charge very high rates (on the order $0.50/kWh and sometimes much more for grid power). Their customers' low ability to pay means they consume very little per month (roughly 5 kWh, enough for the most basic services). The quality of power is often very low given that distribution is at 240 V and power is only provided for a few hours a day over the grid. Despite the poor quality power and high prices, an estimated 90% of the rural population has a lightbulb even though only 15% of the population officially has access to electricity.[1]

The Rural Electricity Entrepreneur (REE) is the dominant model for providing electricity in rural areas of Cambodia. As implied by their name, these are private enterprises undertaken by individuals within a community. They

---

[1] While the government socio-economic survey does not ask about lightbulb ownership (a clear indicator of access to either battery charging or a mini-grid), their statistics do show nearly 50% of the *rural* households owning a television. Thus, the estimate of 90% basic electrification is not unreasonable.

generally have no connection to outside actors, including the government, the main electricity utility, the formal banking sector or the donor community. For the most part, they do not have official licences to operate their businesses and are completely unregulated (changes in this situation are discussed further below).

The target customers are primarily households and some small home-based businesses (though larger ones serving district towns will serve a wider range of shops and productive activities). Average consumption is extremely low (in the range of 5 kWh per month) for the grid customers and four to ten charges per month for those using batteries.

Electricity service by these small entrepreneurs is provided using diesel engines for battery charging or for a combined battery charging and small micro-grid network. There are approximately 1,000 battery chargers and another 600 that run a grid.[2] The technical quality of these operations is quite low and the losses over the thin wires are quite high (reaching 50%)

Start-up capital, as well as capital for improvements to meet licensing requirements, generally comes from sale of prior assets or informal loans at very high interest rates. Tariffs are set to recover costs, though there is not always clear accounting separating battery charging from network based sales. The resulting prices are high (up to $1/kWh and roughly $0.40/charge), however the low consumption means that overall budget outlays per month for customers is within their ability to pay.

There are a number of institutional changes which may start affecting the REE business model. The first is the development of a formalized regulatory mechanism. In theory, the REEs should have a licence to operate from the regulatory authority in Phnom Penh. Some, including large ones in the district towns, have applied for licences from the Electricity Authority of Cambodia. Roughly 20% of the REEs operate at the district town level. However, the ones that are applying for licences appear to be those that are concerned about the other major shift in the electricity sector. Électricité du Cambodge plans to expand its system. It would like to serve a 40 km radius around the provincial capitals and along the planned transmission corridors that will bring power from the neighboring states. Given the relatively dense packing of rural communities in Cambodia, this would include a large proportion of the Cambodian population. The plan would likely be implemented in phases, starting with the largest population centers. Hence a number of REEs in those areas are beginning to upgrade their systems and obtain licences to operate in advance of EDC's expansion. Their licences are only for limited terms and there are no guarantees of renewal. However, the implicit assumption appears to be that the licenced REEs would be

---

[2] The number of battery chargers appears to be more uncertain than that of rural electricity entrepreneurs that run micro-grids (with or without associated battery charging service). The micro-grid operators tend to be larger and more visible. The number of 1,000 for battery chargers is likely a low estimate.

allowed to keep the distribution portion of their business, becoming in effect retailers of EDC electricity.

The case of the REEs in Cambodia is an extremely important one. It is well known that private entrepreneurs using diesel generators are a major force for rural electrification in a number of countries. However, their contribution is generally not counted in official statistics or reported in the literature. Table 4.1 provides a summary of the case.

Table 4.1 Summary of the rural electricity entrepreneur (licensed and unlicensed) (Sources: Enterprise Development Cambodia 2001; World Bank 2003; National Institute of Statistics 2009; Int.Don3 April 2006; Cam.Cons2 December 2005; Cam.NGO.Dom December 2005; Cam.Gov1 December 2005; Cam.Gov2 December 2005; Cam.Site.Rice March 2006; Cam.Ind.REE1 March 2006; Cam.Ind.REE2 March 2006; Cam.Ind.REE3 March 2006; Cam.NGO.Dom March 2006; Cam.Site.Br March 2006; Cam.Site.Ice March 2006)

		Licensed REEs	Unlicensed REEs
DREM parameters	Organization	Entrepreneurs	Small entrepreneur
	Target customers	Medium sized communities	Households and home based businesses
	Technology	Diesel mini-grids	Diesel mini-grids plus battery charging stations
Control variables	Financial: capital	Loans	Informal loans
	Financial: O&M	Cost-recovering tariffs	Cost-recovering tariffs
	Capital cost subsidies	None	None
	Operating cost subsidies	None	None
	Density of customers	High	High
	Remoteness of customers	Low	Low
Outcomes	Policy regime	Favorable	Unfavorable
	Regulatory Regime	Favorable	Unfavorable
	Access	High	High
	Sufficiency	Medium	Low
	Quality	Medium	Low
	Sustainability	High	Medium-high
	Replicability	High	High
Notes on institutional factors	Policy measures	Move to bring all electricity supply under single regulatory and policy framework	
	Regulatory measures	Licensing requirements, protection of licensed distributors	Expansion of EDC jeopardizing REE model over long term but varies by area
	Other		

DREM Parameters

Organization: Individual entrepreneurs

Technology: Diesel generators with electricity provided either through micro and mini-grids or by charging batteries.

Target Customers: The REEs serve entire communities include some productive activities such as shops, welders, etc. The licensed REEs serve larger communities.

Financial Structure (Capital): All capital comes in the form of loans or equity. The smaller REEs, which are all unlicensed, rely on informal loan networks with high interest rates. Capital is recovered through the tariff.

Financial Structure (O&M): All operating and maintenance costs are recovered through the tariff.

Control Variables

Capital Cost Subsidies: There are currently no capital cost subsidies. The Rural Electrification Fund would provide 25% subsidies for qualifying projects.

Operating Cost Subsidies: There are no operating cost subsidies.

Density of Customers: The density of customers is relatively high which is what allows for the mini-grids and what makes it feasible to expect battery charging customers to drop off their batteries regularly.

Remoteness of Customers: The Cambodian rural population is primarily clustered near to district and provincial towns and along roads. However, the road conditions make travel difficult.

Policy Regime: The policy regime is favorable for those REEs that are licensed or seeking licenses since they are participating in the process of regularizing the industry. For unlicensed REEs the policy regime is unfavorable as expansion of EDC poses a threat to their distribution business.

Regulatory Regime: The regulatory regime is favorable for licensed REEs and affords a degree of certainty regarding future business opportunities once EDC expands. For the small, unlicensed, REEs the regulatory regime is less favorable as it requires a licence. This involves a lot of paperwork and significant investment to meet the regulated standards.

Outcomes

Access (HIGH): The REEs are the primary means by which the Cambodian population receives electricity. The licensed REEs provide nearly as much electricity as the EDC, while the unlicensed REEs reach nearly the entire rest of the population.

Sufficiency

Licensed REEs (MEDIUM): Licensed REEs have larger systems and are able to provide more power to the customers.

Unlicensed REEs (LOW): Unlicensed REEs provide a bare minimum of power, usually only a few kWh per month to each household and only in the evenings. However, they are able to provide power throughout the day for a few customers that are willing and able to pay (e.g. productive activities). It is unclear how much unmet demand there is given the economic situation of many REE customers (i.e. they may not be able to afford many more appliances or the electricity to run them, even if it was available).

Quality

Licensed REEs (MEDIUM): Licensed REEs have to meet certain standards in order to obtain and maintain their license.

Unlicensed REEs (LOW): The electricity quality is very low for the unlicensed REEs. There are high line losses and the power quality is poor.

Sustainability

Licensed REEs (HIGH): The costs of service are fully recovered by the electricity entrepreneurs. Their licenses give them a degree of certainty regarding the legitimacy of their business.

Unlicensed REEs (MEDIUM-HIGH): Like the licensed REEs, these entrepreneurs are able to recoup their costs through the tariffs they charge. However, without licenses, they are run the risk of being shutdown for operating illegally. The investment necessary to meet the licensing standards may be too high for some REEs.

Replicability (HIGH): The rural electricity entrepreneur business model has been widely replicated and exists in every region of Cambodia. Furthermore, evidence indicates that similar business models exist in a number of other countries, particularly in regions not served by either centralized utilities or government programs.

Solar Home Systems and PV for NGO Projects

Another DREM to consider is the cash market for solar systems. Khmer Solar has two markets that it serves.[3] The first is a home-based market and the second is for NGO projects and NGO offices in the field. Solar home systems are sold out of

---

[3] There is at least one other company in Cambodia supply solar systems, however its target customer is the government (e.g. for military installations).

Khmer Solar's main office in Phnom Penh and a retail store in Battambang (approximately 4 h NW of Phnom Penh). The most popular system that Khmer Solar sells to households is an 85 W system for approximately US$750 (roughly two times the per capita GDP). This includes the panel plus battery, all of the balance of system, operation of the Battambang store, installation, etc. However, programmatic costs incurred by the Phnom Penh office to support the Battambang store are covered by higher costs charged to Phnom Penh and NGO customers. The equivalent tariff (incorporating capital costs and accounting for replacement parts) is 78 c/kWh. There are three credit systems offered (6 months, 1 year and a harvest time credit). The credit program requires 40% downpayment and then the remaining paid in 6 months (at no interest), 12 months (at 1%/month interest). However, the high costs, even with the credit system, limit sales to those on the higher end of the rural income scale. These are also consumers who have higher consumption than is usually served by battery chargers or even the small micro-grids (roughly 10 kWh per month rather than 3–5 kWh per month). Systems also come with a 1-year warranty (after which repairs must be paid for on a cash basis). The customers of these systems are richer rural residents who purchase them for their convenience (no need to transport batteries) and ability to meet higher power demands. The market is still relatively small (200 units sold in 2005). Three obstacles stand in the way of expanding the market. First is the economic situation of most rural residents, which precludes them from having the money necessary to purchase such a system. Battambang was chosen for the first retail store because that province's economy is better than most of the rest and the population is somewhat richer. Second, would be the logistics of expansion. Without opening new stores in other provincial towns, installation, maintenance and credit payment collection is made more difficult. Third, solar home systems in Cambodia are relatively expensive, in part due to a 50% import tariff on the solar panels. Negotiations to reduce or remove the tariff have been unsuccessful.

Interviews with three solar home system customers indicate similar reasons among all three for purchase of the solar home system.[4] Previously all three were reliant on battery charging services which were expensive and entailed regular trips to the charger. The solar home system saves them both time and money. Environmental factors did not a play a role in their decision-making.

In addition to serving the household market, Khmer Solar sells systems to various NGOs working in rural areas and provides solar systems as part of an international NGO project to build rural schools. These are essentially the same systems that are sold to the households. The only differences are in the credit programs (non-existent for the NGO projects) and the price, which is slightly higher for the NGO projects. The higher revenue from the NGOs is used to fund

---

[4] The three solar home system customers were: (a) a rice farmer with above-average landholdings; (b) an elderly couple that had lived with their children in the United States for many years and (c) a local entrepreneur who collected recyclables and ran a lottery system for which he needed electricity to run a transmission system to his lottery ticket sellers in the countryside stores.

the operation of the main office of Khmer Solar in Phnom Penh, effectively resulting in a small cross-subsidy between the international NGOs purchasing systems for their field operations and the households purchasing systems for their own use.

The Khmer Solar business model shows strong financial sustainability. Overall, the costs of the system are covered by the purchasers. There is a slight reduction in costs to the rural households since the programmatic costs of the Phnom Penh office are covered by NGO projects and sales to richer urban customers in Phnom Penh. However, the rural customers are covering the costs of the systems themselves. There are trained technicians and the customers are aware that there are ongoing costs beyond the first year of warranty. They are already accustomed to having to replace batteries from when they used the service of battery chargers.

However, further expansion of this sales model faces a few hurdles. First, there are high import duties on the solar panels (50%) which prevent further cost reductions. Since the solar panels are roughly half the cost of the entire system, even with elimination of this import duty, the costs of the systems would remain above that affordable by most households. Second, and perhaps more importantly, there is a logistical problem in expansion. Collecting payments and servicing systems in remote regions poses a problem. At the moment, sales are limited to the Battambang region where there is a relatively high density rural population living within a close distance to the city. Table 4.2 provides a summary of the case.

DREM Parameters

Organization: Local company

Technology: Solar home systems with battery.

Target Customers: Wealthier rural households and non-governmental organizations working in rural areas.

Financial Structure (Capital): This is a cash market where capital is recovered by the selling price. There is a slight cross-subsidy from the NGOs and richer urban customers to the rural households since programmatic costs of the Phnom Penh office are not included in the rural household price.

Financial Structure (O&M): The cost of the system includes a 1 year warranty, after which costs are borne by the customers.

Control Variables

Capital Cost Subsidies: There are currently no capital cost subsidies. On the contrary, there is a 50% import duty on the solar panels. The Rural Electrification Fund would provide 25% subsidies for qualifying projects.

Table 4.2 Summary of Khmer solar model (rural home and NGO sales) (Sources: Cam.Cons December 2005; Cam.Gov1 December 2005; Cam.Gov2 December 2005; Cam.HH1 March 2006; Cam.HH2 March 2006; Cam.HH3 March 2006; Cam.Ind.KS1 March 2006; Cam.Ind.KS2 March 2006; Cam.NGO.Dom March 2006)

		Solar home systems	NGO PV
DREM parameters	Organization	Dealer	National/international NGO
	Target customers	Households	Community structures
	Technology	PV SHS	PV
	Financial: capital	Cash market	Donor
	Financial: O&M	Warranty plus cash	Warranty plus cash
Control variables	Capital cost subsidies	Low	None
	Operating cost subsidies	None	None
	Density of customers	High	Low
	Remoteness of customers	Low	High
	Policy regime	Neutral	Neutral
	Regulatory Regime	Neutral	Neutral
Outcomes	Access	Low	Low
	Sufficiency	High	High
	Quality	High	High
	Sustainability	Medium	Medium
	Replicability	Medium	Medium
Notes on institutional factors	Policy measures	Lack of agreement on eliminating tax keeping prices inflated	Lack of agreement on eliminating tax keeping prices inflated
	Regulatory measures		
	Other	Sustainability will depend on Khmer Solar's continued operation	Sustainability will depend on Khmer Solar's continued operation

Operating Cost Subsidies: There are no operating cost subsidies.

Density of Customers: The density of customers is mixed. Some purchase solar system even when they can use a battery charger due to the cost and convenience. Others are in less dense rural areas.

Remoteness of Customers: The Cambodian rural population is primarily clustered near to district and provincial towns and along roads. However, the road conditions make travel difficult.

Policy Regime: The policy regime is neutral regarding solar systems, with the exception of the import duty which is unfavorable.

Regulatory Regime: There are no regulations governing the sale, distribution and use of solar home systems.

Outcomes

Access (LOW): The market for solar home systems remains small (only a couple of hundred systems were sold in 2005). Sales for households are only centered around Phnom Penh and Battambang.

Sufficiency (HIGH): One of the advantages cited by the customers interviewed was the ability to consume more electricity easily than when they used the battery charging service. The modularity also allows users that can afford to do so an easy method to add more capacity (e.g. the rural lottery system operator who used one system for his household and another for power his transmitter).

Quality (HIGH): The systems are reliable and there aren't the voltage problems associated with transmission over a poor distribution system.

Sustainability (MEDIUM): The cost-recovering financial structure and the willingness and ability to pay of the customers indicates that this model is sustainable. However, there has not been enough time since its inception to make a definitive statement. Furthermore, expansion of the EDC grid will mean that customers that might benefit now from the technology will be better off with the grid in the future and no longer be customers.

Replicability (MEDIUM): There is a functioning business model and a clear willingness and ability to pay among a segment of the rural Cambodian population. However, there are logistical difficulties in serving certain parts of the country that are more remote. Certainly this cannot be done solely from Phnom Penh and Battambang.

Biomass Cooperative

Given the relatively long time-frame of EDC's master plan for grid expansion, a number of other activities are occurring to expand the use of distributed generation in rural areas. Small and Medium Enterprises Cambodia, an NGO, has set-up a rural energy cooperative based on biomass gasification technology imported from India. The cooperative members (currently 81 households and a couple of stores) not only receive electricity through a mini-grid, they are also able to supplement their income by selling the biomass to fuel the gasification system. While the capital costs for the system were donated, users do pay a full cost-recovering tariff (including depreciation), and plans for future expansion of this system and for further systems would reduce the grant component through loans and increased equity from cooperative members. The current tariff is roughly 38 c/kWh. The fact that the cooperative has begun by charging a cost-recovering tariff is a strong indication that this model could be sustainable even as grant funds are replaced by loans and equity.[5]

---

[5] The cooperative leadership was also able to pass a rate hike when it was realized the initial rate was set too low. This can be compared to the dispute over the tariff in the Japanese PV/Hydro project discussed below.

With money from the Canadian International Development Agency and logistical support from a Cambodian NGO (SME Cambodia), a biomass gasification system was installed in a village near Battambang in the Northeast. The cooperative has two related functions. First, it oversees a tree farm that grows the fast-growing leucaena tree. Each member of the cooperative has a plot of land in the farm to grow trees, which are then sold to the cooperative, providing them with some cash income. The second aspect of the cooperative is the use of this biomass in a gasification system to generate electricity for the roughly 80 households that are members of the cooperative.

The biomass gasification system is from India and cost roughly US$28,000 to install and purchase. It is a 25 horsepower system with a peak output of 9 kW (the rated 12 kW peak power output cannot be reached due to necessary diesel engine modifications). Similar to the micro-grids run by the REEs, the system uses three phases to reach the households (and a few stores) plus has a battery charging service. The system produces about 1,200 kWh per month with roughly half from battery charging. The fuel comes from 8 ha of tree plantation.

The financial model of the cooperative stresses cost recovery (despite the initial donation of the capital funds for the initial system). All operating costs are covered by the power tariff. Furthermore, the tariff is sufficient to allow for investment in a future expansion of the system by leveraging funds through the REF. The tariffs at the time of the site visit were 1,500 Riel/kWh for grid power and 1,300 Riel/50 Ah for battery charging. The grid connected price is roughly US$0.38/kWh., which is more than EDC but less than the average REE tariff. This also does not account for the fact that cooperative members receive 60 Riel/kg which amounts to roughly $200/ha/year. A household consuming roughly 10 kWh/month would have annual electricity bills of US$45 but revenue from the crop of $20, significantly offsetting their electricity costs (not accounting for a one-time connection fee of US$12.5).

Future plans for expansion of system and new systems would reduce the donor grant component and replace with higher community contributions and loans. This will be necessary in order to tap into funds from the Rural Electrification Fund which requires 25% equity, 25% subsidy and 50% loans. Expanded plantations (an additional 15 ha) would provide enough biomass for a 30 kW system. Table 4.3 provides a summary of the case.

DREM Parameters

Organization: Cooperative established with the aid of a domestic non-governmental organization.

Technology: Indian biomass gasification system and modified diesel engine.

Target Customers: Entire community. Cooperative members provide biomass feedstock as well as obtain electricity.

Financial Structure (Capital): Initial capital provided largely by the Canadian International Development Agency. Tariff covers capital replacement and expansion costs.

Financial Structure (O&M): Cost recovering tariff.

Table 4.3 Summary of biomass cooperative model (Sources: Cam.NGO.Dom December 2005; Cam.Site.BaBio March 2006; Cam.NGO.Dom March 2006)

		Biomass cooperative
DREM parameters	Organization	Community Based Coop/National NGO
	Target customers	Households
	Technology	Biomass gasification mini-grid
	Financial: capital	Donor plus community
	Financial: O&M	Cost-recovering tariff
Control variables	Capital cost subsidies	Medium
	Operating cost subsidies	None
	Density of customers	High
	Remoteness of customers	Low
	Policy regime	Neutral
	Regulatory regime	Neutral
Outcomes	Access	Low
	Sufficiency	Medium
	Quality	Medium
	Sustainability	Medium
	Replicability	Medium
Notes on institutional factors	Policy measures	
	Regulatory measures	Expansion of EDC jeopardizing Coop model over long term
	Other	Replicability will depend on emergence of financing structures other than international donor grants

Control Variables

Capital Cost Subsidies: There are no capital cost subsidies provided by the government. The REF program would provide a 25% subsidy for future expansion.

Operating Cost Subsidies: There are no operating cost subsidies.

Density of Customers: The customer base is dense enough to create a micro-grid and to have some customers bring batteries.

Remoteness of Customers: The community is not far from other communities and not far from one of the major routes.

Policy Regime: There are no policies in place to either encourage of discourage these types of projects.

Regulatory Regime: While there are rules governing micro-grids and licensing requirements, generally the regime is somewhat neutral as it recognizes that there are a lot of activities that occur outside the regime that are necessary.

Outcomes

Access (LOW): Thus far, this model has been attempted in only one village.

Sufficiency (MEDIUM): Even with its capacity reduced to 9 kW, the system is able to provide more than the energy usually consumed by battery-charging households or even other micro-grid customers. However, the desire to expand the system indicates there is unmet demand (either within existing cooperative member households or from nearby households that are not cooperative members).

Quality (MEDIUM): The major problem has been that the system was purchased from India. While downtime has been minimal, repairs have been made more difficult because the system uses parts that are not standard sizes for Cambodia (which does not conform to the British standard for items such as screws or nuts).

Sustainability (MEDIUM): The cooperative is still young but has shown an ability to make difficult decisions (e.g. raise tariffs) and has been able to continue despite a few breakdowns in the system.

Replicability (MEDIUM): There are three obstacles to similar cooperatives being established elsewhere in Cambodia. First, is the need for some person or organization to take the initiative. Second, the initial investment costs must be covered, either through a loan (likely to be an informal one at high rates) or through a donation. These two obstacles were overcome in this case through the involvement of the national NGO that had the capacity to organize and to find funding. It is not clear who would take up that role for future cooperatives. The third is obtaining the technology from suppliers that would require less modification and pose fewer maintenance problems.

Japanese PV/Hydro Project

The final mini-grid based project examined is a relatively large donor project based on a combination of micro-hydro and solar photovoltaic technologies. The end of the conflict in Cambodia has resulted in an influx of donor aid as part of a post-conflict reconstruction and general development effort. In addition to programmatic aid, such as the development of the Rural Electrification Fund, aid for specific projects has come from individual donor governments. The biomass cooperative discussed above was started with the assistance of Canadian Government funds. However, primary implementation and decision-making was local. By contrast, the Japanese also set up an energy project in a rural area but with all funds coming directly from the Japanese government and administered centrally by the Cambodian Ministry of Mines and Energy. This type of donor funding is more reminiscent of the types of projects that used to occur more frequently in the seventies and eighties. Funded by the Japanese government, this $3 million project consists of a combination PV/hydro mini-grid and five satellite PV battery charging stations. Total capacity is 100 kW and roughly 410 households are served by the mini-grid. Electricity is provided from 6–9 pm with a peak load of 40 kW. The project, however, has run into problems. First, is a dispute over the tariff, currently set at 16 c/kWh. This tariff is too low to recover basic operations and maintenance costs. An increase to 25 c/kWh is being proposed but thus far been rejected by the local community, who are only willing to pay a higher tariff if they get power for longer periods in the day and more power for productive activities. Second, is the fact that at the moment the mini-grid users are not getting any power at all due to technical problems. The PV system feeding the micro-grid broke down almost immediately. The micro-hydro system continued working for a while, but also broke down in February or March of 2006. Funds were never provided for equipment replacement and, as mentioned above, the tariff was insufficient to cover such repairs. This situation created a catch-22. There was no money to fix the system and tariffs couldn't be raised until the system was fixed and customers could get reliable power. The problem was made even worse a few weeks before the site visit when the hydro system also broke down leaving only the satellite PV battery charging stations functional. With limited revenue coming in the workforce was reduced to one person. Table 4.4 provides a summary of the case. As a result, the mini-grid portion of the project is not operational and there is no definite timeframe or process for resolving the problem.

Table 4.4 Summary of Japanese donated PV/hydro model (Source: Cam.Cons December 2005; Cam.Site.PVH March 2006)

		Japanese PV/hydro
DREM parameters	Organization	International donor/national government
	Target customers	Households
	Technology	PV/hydro mini-grid plus battery charging stations
	Financial: capital	Donor
	Financial: O&M	Insufficient tariff
Control variables	Capital cost subsidies	High
	Operating cost subsidies	None
	Density of customers	Medium
	Remoteness of Customers	Medium
	Policy regime	Favorable
	Regulatory regime	Neutral
Outcomes	Access	Low
	Sufficiency	Medium
	Quality	Low
	Sustainability	Low
	Replicability	Low
Notes on institutional Factors	Policy measures	
	Regulatory measures	
	Other	

DREM Parameters

Organization: International Donor and National Government
Technology: Mini-Hydro and Photovoltaic Mini-Grid plus Photovoltaic Battery Charging Stations
Target Customers: Households in a community
Financial Structure (Capital): All capital donated by Japanese government. No provision in the tariff for capital replacement or expansion.
Financial Structure (O&M): Non-cost recovering tariff and no provision for major repairs.

Control Variables

Capital Cost Subsidies: All the capital was provided as a donation by the Japanese Government.
Operating Cost Subsidies: There are no operating cost subsidies.
Density of Customers: The main community is dense enough to be served by micro-grid while the solar charging stations serves those that are further removed from the village.
Remoteness of Customers: The village was further from the main road than many other villages in the area.
Policy Regime: The Cambodian government is very supportive of bi-lateral and other types of development projects.
Regulatory Regime: As with similar projects, the regulatory authority is not concentrating on these types of operations.

Outcomes

Access (LOW): Only one community is served by this system (roughly 400 households)
Sufficiency (MEDIUM): The system did provide enough power in the evenings for most rural household uses. However, the complaints that power was not provided for productive activities indicates that a system with higher capacity would have been more suitable.
Quality (LOW): Power was only provided for part of the day and then the failure in the PV station and the hydro station has meant there is no power going over the mini-grid at all.
Sustainability (LOW): This model has already partially failed since the technology had a major problem early on and there is no money to fix it. Furthermore, the institutional mechanisms are not in place to properly run it.
Replicability (LOW): This model is heavily dependent upon the donor funds to build it and its performance has been poor.

Key Findings and Conclusions

The lack of strong centralized institutions and resources for rural electrification have led to Cambodia's rural electrification being dominated by small entrepreneurs running diesel-based battery charging stations and small mini-grids. The change in institutional setting that occurred in Cambodia also provides valuable lessons. In the post-war era there has been an influx of investment and donor aid that has led to individual projects, programmatic aid and development of a Rural Electrification Fund. At the same time, in 2001 a new electricity law created the Electricity Authority of Cambodia, an independent regulator. The EAC has established licensing procedures for all levels of the electricity industry. These changes have resulted in both successful projects (e.g. the biomass cooperative) and unsuccessful projects (e.g. the Japanese project) and a more uncertain institutional environment for the REEs.

Électricité du Cambodge, the government-owned utility of Cambodia, is also in the process of expanding its system. At the time of this research, EDC only served roughly 9% of the population, limited to the national capital and some provincial capitals. However, its master plan envisages grid coverage of all major population centers and extension along the main roads from those centers. This would bring it into areas currently served by REEs, creating greater uncertainty for the rural electricity. The result is a gradual formalization of the sector. Many of the REEs in the larger towns (mainly district towns) have already sought licenses from EAC for their service territory. Similarly some of the small REEs are upgrading their distribution lines and seeking licenses in order to be considered as distributors when the EDC grid reaches their area. Licenses are being granted (EAC had issued a total of around 140 consolidated licenses by 2006 and is now up to 267 licenses), however the REEs are being granted conditional and short-term licenses (1–5 years). The conditions are that the REE has to upgrade their distribution lines to meet certain technical requirements. There is, however, no guarantee of renewal of the distribution portion of their licenses once EDC's grid arrives and no provision for compensation for their stranded investment in generation equipment. The licensing procedures and requirements appear geared towards larger operators and could impose severe burdens on the smaller REEs who now find themselves technically operating illegally.

Cambodia's rural electrification is also interesting because it demonstrates on a large scale that consumers in rural areas have a high willingness to pay for electricity, even if they can only consume a little electricity per month. This allows the diesel based rural electricity entrepreneurs and the Khmer Solar company to enjoy full cost-recovery and be successful. Both models, though different, are cost-recovering models that are sustainable and replicable. The rates they must charge in order to recover their costs are quite high in comparison to either the ability to pay of most rural customers or the rates paid by urban customers. They are also quite high in comparison to what is often paid in other countries with subsidized electrification programs. However, their success shows that there is a willingness to pay among rural customers and that even within the rural community there are differing customer classes. The PV systems are marketed to slightly richer rural customers who can afford the upfront costs and who are seeking slightly higher energy consumption. The REEs make their business model work by segmenting the market, serving most customers with only a modest amount of electricity in the evenings, recharging batteries, and serving a small number of customers with higher ability to pay throughout the day. In the case of the rural entrepreneurs, they are also responsible for the vast majority of electricity access in the country. The large number of mini-grids also shows that the organizational difficulties associated with operating such systems can be overcome.

The need for cost recovery and for the establishment of institutional mechanisms for sustainability is further reinforced by the failure of the Japanese PV/hydro project. This DREM also illustrates the difficulties encountered with top-down centralized models (as opposed to the REEs and PV solar home system sales). However, the Japanese project also demonstrates that high subsidies are also not a guarantee of high impact in terms of immediate changes in electricity access.

Of the other two models, the biomass cooperative attempts to replicate a fully cost-recovering operation as closely as possible, despite the initial donation of equipment costs. This may explain, in part, its ability to handle early setbacks and make adjustments in tariffs. This is in contrast to the Japanese donated system. Another important difference is in the participation of the local community. In one case, a cooperative was established and the community engaged in decision-making. In the other case, a community group was formed, but it seems to have been done after the fact and have a much more adversarial relationship with the operators of the system. The Japanese PV/Hydro project also shows that high subsidies don't necessarily have a high impact on immediate improvements in electricity access. The Government of Cambodia's own rural renewable electrification policy explicitly recognizes that the advantages of renewable energy come from its competitive pricing vis-à-vis diesel based generation (MIME 2003).

The Cambodian case, like the Brazilian case, shows how much of an impact a change in the institutional environment can have on distributed rural electrification models. In this case, it is the establishment of the regulatory authority and the future expansion plans of the centralized monopoly that is creating pressures on DREMs, particularly the rural electricity entrepreneurs close to and in district towns (Int. Don3 April 2006). Fear of a complete takeover of their distribution area without compensation is leading them to become formalized players. They are obtaining licenses to at least protect the distribution side of their business into the future.

The Cambodian case also mirrors the Brazil case in showing both the role of non-household electrification and the differing outcomes that can result from installing renewable energy sources. Part of the market segmentation that makes the REEs successful is the ability to serve small commercial customers during the day. These consumers provide another revenue stream that helps their profitability. This is another area that the Government of Cambodia has recognized is important to the prospects of rural electrification. Their renewable energy for rural electrification policy calls for programs and projects to be integrated into development planning and to serve productive activities and socially beneficial activities (e.g. schools) in addition to households (MIME 2003). This need for integrated rural energy planning that accounts for diverse energy needs beyond household electrification and recognizes the potential for decentralized energy is also recognized by at least parts of the international donor community (Global Environment Facility undated).

It is also clear that the renewable energy projects in Cambodia were not always chosen specifically for climate change reasons. The solar home system customers, for example, view them as a more convenient and economically attractive alternative to diesel battery charging. The biomass cooperative allowed the local community to be involved in fuel supply. While there has been some interest in converting the diesel based REEs into renewable energy based battery chargers, the motivations listed in documents have more to do with cost than with the environment. However, there was no evidence at the time of this research that this has taken off as an option (Brun et al. 2002). On the other hand, due to its mandate, the Global Environment Facility sees rural renewable energy projects as having both a development and environmental benefit (Global Environment Facility undated).

References

Brun J-M, Mahe J-P et al (2002) Photo-voltaic market development in Cambodia: potential involvement of ESCO and batteries/appliances traders. Kosan Engineering for the United Nations Development Program, Phnom Penh, Cambodia, p 82

Cam.Cons (December 2005) Author interview with renewable energy consultant. Phnom Penh, Cambodia

Cam.Cons2 (December 2005) Author interview with renewable energy consultant. Phnom Penh, Cambodia

Cam.Gov1 (December 2005) Author interview with electricity authority of Cambodia. Phnom Penh, Cambodia

Cam.Gov2 (December 2005) Author interview with Ministry of Industry, Mines and Energy. Phnom Penh, Cambodia

Cam.HH1 (March 2006) Author interview with household. Battambang Province, Cambodia

Cam.HH2 (March 2006) Author interview with household. Battambang Province, Cambodia

Cam.HH3 (March 2006) Author interview with household. Battambang Province, Cambodia

Cam.Ind.KS1 (March 2006) Author first interview with renewable energy enterprise. Phnom Penh, Cambodia

Cam.Ind.KS2 (March 2006) Author second interview with renewable energy enterprise. Phnom Penh, Cambodia

Cam.Ind.REE1 (March, 2006) Author interview with rural electricity entrepreneur. Sneoung, Cambodia

Cam.Ind.REE2 (March 2006) Author interview with rural electricity entrepreneur. Mongkol Borei, Cambodia

Cam.Ind.REE3 (March 2006) Author interview with rural electricity entrepreneur. Phnom Toch, Cambodia

Cam.NGO.Dom (December 2005) Author first interview with domestic non-governmental organization. Phnom Penh, Cambodia

Cam.NGO.Dom (March 2006) Author second interview with domestic non-governmental organization. Phnom Penh, Cambodia

Cam.Site.BaBio (March 2006) Author site visit, biomass gasification facility. Battambang Province, Cambodia

Cam.Site.Br (March 2006) Author site visit to rural brick factory. Battambang, Cambodia

Cam.Site.Ice (March 2006) Author site visit to rural ice factory. Banteay Meanchey, Cambodia

Cam.Site.PVH (March 2006) Author site visit to donor PV/hydro project. Kampong Cham, Cambodia

Cam.Site.Rice (March 2006) Author site visit, rice mill. Battambang Province, Cambodia

Enterprise Development Cambodia (2001) Survey of 45 Cambodian rural electricity enterprises. Enterprise Development Cambodia, Phnom Penh, p 108

Global Environment Facility (undated) Fostering modern energy systems in rural Mekong. Washington, DC, p 101

Int.Don3 (April 2006) Author interview with international donor. Washington, DC

Int.Don4 (January 2006) Author interview with international donor. Washington, DC

Kingdom of Cambodia (2001) Electricity law of the Kingdom of Cambodia. Royal Decree No. NS/RKM/0201/03

MIME (2003) National policy on renewable energy-based rural electrification. Ministry of Industry, Mines and Energy (MIME), The Royal Government of Cambodia, Phnom Penh, Cambodia

National Institute of Statistics (2009) Housing conditions 2007. Government of Cambodia, Ministry of Planning, National Institute of Statistics, Phnom Penh, Cambodia

National Institute of Statistics (undated) Cambodia socio-economic survey 2004 (Summary). Government of Cambodia, Ministry of Planning, National Institute of Statistics, Phnom Penh, Cambodia

World Bank (2003) Project appraisal document on a proposed credit in the amount of SDR 27.9 million (US$40 million equivalent) and a proposed GEF grant of US$5.75 million to the Kingdom of Cambodia for the rural electrification and transmission project. Washington, International Bank for Reconstruction and Development, Energy and Mining Sector Unit, SE Asia and Mongolia Country Unit, East Asia and Pacific Region, p 146

Chapter 5
Distributed Rural Electrification in China

Keywords China • Hydropower • State government

Introduction

The sharp reduction in unelectrified households in China, particularly over the last 20 years has been remarkable. From hundreds of millions of people without access, China had perhaps 15–20 million people without access to electricity at the time of this research and 8 million today (IEA 2009). The central government has even established criteria by which counties are judged to be "electrified," including minimum levels of consumption.[1]

The history of rural electrification in China is usually divided into anywhere from three to five distinct stages, depending on the analyst. For this analysis, the rural electrification history of China will be divided into three stages (Pan et al. 2006). The first stage, roughly from 1949 to 1978 was one of steady expansion of the system. The goal was economic development and the role of the state in planning was strong. However, "ownership" was effectively shared between the centre and the local authorities. During this phase a number of favorable policies were put in place regarding tariffs, subsidies and the use of electricity revenues to encourage further development of the system. By the end of this stage rural access rates had reached 63%.

The second stage, from 1979 to 1998, was one of significantly more rapid expansion. While keeping a focus on the expanding industrial base of the country, the mandate for electrification was expanded to include households more directly. Pilot counties were prioritized for electrification in a series of programs funded by

*Information on Distributed Rural Electrification Models in China comes from a combination of secondary sources and primary interviews conducted in China (March 2005, May/June 2005, December 2005) and Washington, DC (2005 and 2006).

[1] Requirements include 90% of the rural households have access to electricity for lighting, broadcasting, TV round the year and cooking seasonally, and for irrigation, food processing and TVEs. On average, per capita electricity consumption is around 200 kwh and generation capacity at 100 W.c. Details are given in GB (National Standards)/T 15659–1955.

the state. While central support remained strong, control shifted to local actors as there was a transition to a market system in the larger economy. During this time rural access rates reached 99%.

The final phase, from 1999 to the present, is one in which aspects of the system underwent a process of reform, consolidation and upgrades. Increasingly rural and urban electricity markets were integrated as the centralized grid reached into more rural areas, including areas already being served by isolated systems. This stage included a push for fuel substitution (particularly the partially successful attempts to shut down small coal plants), restructuring of the broader electricity sector, increased investments in rural grids, and a focus on reducing disparities between eastern and western regions of the country. While the result of much of this activity was improvements in electricity service in numerous areas, it was not without its conflicts, particularly between the central organizations and local actors over ownership of assets.

Each of these three stages was marked by a shift in institutions, policies, incentives and industry structure. The efforts on rural electrification were manifest through a shifting set of organizations and institutions, including the grid companies, state banks, ministries, the development commission, local governments, local utilities, price bureaus and the private sector. However, at times this meant that the lines of authority, as well as property rights and ownership were not always well defined. There is also a marked shift in the focus of electrification efforts with agriculture, industry and households having a different emphasis at different times, depending on the larger economic and political forces at work in the country, particularly as the country moved to a market based economy and state-local relations were renegotiated. However, despite times of uncertainty and varying degrees of effort on rural electrification, China does stand out as having had remarkable success in electrifying its rural population. This is particularly true during the two and a half decades post-reform when hundreds of millions of rural Chinese obtained access and the total unelectrified population dropped to less than 1% of the population (though this still means some 20 million people, give or take, are still without electricity) (Rufin et al. 2003; Tong 2004; Pan et al. 2006; Barnes 2007 and references therein).

As with much of the Chinese economy in the post revolution years, rural electrification efforts were focused on solving the macro-level problems of the fledgling nation. This meant a focus initially on meeting the agricultural needs of the population and then later to provide energy for the expanding industrial basis in rural areas. Emphasis was placed on supplying agricultural producers and, later on, industry. This can be seen in the share of electricity consumption by different sectors. In 1976, households accounted for 12% of consumption, while irrigation and agricultural processing accounted for 41%. Industrial uses (at the county and village level) accounted for the highest share with 47% of consumption (Tong 2004).

In order to strengthen and broaden the economic base, as well as to deal with severe regional inequities between the east and west of China, there was a pronounced emphasis on rural electrification during the eighties and nineties. Significantly, this included a focus on households that did not exist before (Pan et al. 2006). Programs were introduced and implemented at the national level to create criteria for rural counties to be considered electrified and significant resources were allocated to the task. The first 100 counties program was quickly followed by 200,

300 and 500 county programs over the next decade. Total allocations from the state for these programs amounted to 70 billion RMB (Tong 2004). While some of these funds went towards expansion of the centrally owned and operated grids systems, significant amounts were in support of the small hydropower program.

As a result of these programs and a shift in China's economic structure over the same time period, the customer pattern had shifted considerably by 2001. Households now accounted for 25% of the consumption while county and village industry accounted for 36% and 22% respectively. Agricultural uses and irrigation had dropped below 10% (Tong 2004).

Box 1 Rural Electrification Counties Criteria

As part of the National Primary Rural Electrification County Program that electrified first 100 counties and then had further 200 and 300 county phases, a set of criteria were established to certify that the counties were considered "electrified." The criteria included:

Requirements include 90% of the rural households have access to electricity for lighting, broadcasting, TV round the year and cooking seasonally, and for irrigation, food processing and TVEs. On average, per capita electricity consumption is around 200 kwh and generation capacity at 100 W.c. Details are given in GB (National Standards)/T 15659–1955

The Institutional Context

The institutional context for rural electrification in China has changed significantly over the decades since the establishment of the current state in 1949. This has included not only changes in the ministries responsible for various aspects of rural electrification, but also the major changes that swept through the Chinese economic (and policy-making) system in the transition of the late seventies and eighties and continuing on to today. In the period up to 1979, the Chinese economic and political system was highly centralized. Decisions and plans came from the center and were implemented locally either by local branches of centralized organizations or by the local governments (themselves mainly a local branch of the central government in Beijing). The energy sector was no exception, with control of the sector in the hands of the State Planning Commission, the various ministries for water, energy, oil, the central bank, etc. and, in the case of rural energy, often implemented by their local agents (Lieberthal and Oksenberg 1986; Steinfeld 1998; Oi 1999; Andrews-Speed 2004).

The structure of China's economic and, indeed, political system has changed significantly over the last 25 years, starting with the reforms of 1979. The changes in the economic system allowed for new entrants into a variety of economic sectors, the ability to attract private investment, unprecedented economic growth that has provided significant public funds for rural infrastructure and a host of other benefits. At the same time, it has significantly changed the decision-making system in a number of ways. First, the economic reforms were accompanied by decentralization efforts, sometimes creating new obligations on local authorities without accompanying resources. This has lead them to find other means of obtaining those resources and also to resist central efforts to take those resources away at times (Lieberthal and Oksenberg 1986; Oi 1999; Andrews-Speed 2004; Tong 2004).

Second, the centralized organizations that dominated energy policy-making were restructured and even disbanded in some cases. The energy portfolio was separated from the Ministry of Water and Energy, brought back into the ministry, and ultimately had a number of functions spun off as state owned enterprises (some of which were partially privatized). As a result, the Ministry of Water became its own entity, responsible only for water. A new regulatory authority for energy was created and the highest level policy-making body, the State Planning Council became the National Development and Reform Commission, expanding its powers, including on energy planning (Lieberthal and Oksenberg 1986; Andrews-Speed 2004; IEA 2006).

One major impact that all of this restructuring (both the economic reforms and the decentralization efforts) has had is increasing ambiguities regarding issues such as decision-making authority and ownership of assets which have been exploited by various actors. Local authorities have been able to modify, add or ignore rules to suit their purposes, including on electricity pricing and the use of inefficient equipment. They have also been able to manipulate the system so that local leaders have become owners of assets.

In analyzing the changes in China's policy formulation and implementation in the post-reform era, particularly in the energy field, Lieberthal has characterized it as "fragmented authoritarianism." While he was applying this to decision making in the larger energy sector, particularly for the oil and gas industry and the large centralized electricity infrastructure, it applies to the rural electricity sector as well. Fragmented authoritarianism refers to the fact that China has maintained an authoritarian system in terms of the discipline expected of those at the bottom of the hierarchical structure and the lack of certain freedoms while at the same time that responsibilities and lines of command have become reassigned. This creates an exceedingly complex network of decision making structures in which some decisions have been decentralized geographically (i.e. to sub-national levels of government) and others have been decentralized according to function (i.e. by sector). The result is that decision-making has become a function not only of law but also of idiosyncratic negotiations at multiple levels and between multiple actors (Lieberthal and Oksenberg 1986; Andrews-Speed 2004). It has also allowed agents to manipulate the system for their own benefit.

This "fragmented authoritarianism" can be seen in the rural electrification sector. It is part of the reason that ownership structures are unclear for some small-hydro power systems and why some private systems are owned by former local bureaucrats and local Ministry of Water Resources officials. It is also why prices vary widely despite the existence of price schedules. Interestingly, it could also partially explain the ability of the national grids to take over some rural systems *and* the ability of those local systems to fight to regain at least partial control.

Major Actors in the Energy Sector

There are a number of actors at all levels that have played an important role in rural electrification in China. The following is a list of the key organizations and their roles:

State Commissions: The National Development and Reform Commission, and its predecessor the State Planning Commission, is broadly responsible for setting policy and spending priorities of the central government. It contains a number of sub-bodies responsible for energy planning. More recently (and subsequent to the fieldwork for this study), a new body has been formed called the National Energy Leading Group (NELG). It is intended as a high level body for energy planning with responsibilities across a range of energy issues. Prior to the creation of the NDRC, there was also a State Economic and Trade Commission. While the SPC was responsible for long-range planning, policy formulation and investments, the SETC was responsible for creating annual plans and implementing them (Andrews-Speed 2004). When the SPC became the NDRC (and effectively took over all responsibility for the economic reform process), the SETC was merged with another ministry responsible for trade and became the Ministry of Commerce.

Ministry of Energy: A defunct ministry, it was at one point in time the state body through which all energy infrastructure was planned and built. For a long time it was part of a ministry that was responsible for both water and energy. Eventually, only the water side of the ministry remained as a ministry (see below). Energy was its own independent ministry twice since 1979 before being disbanded. State electricity assets were corporatized and a regulatory authority created. State electricity assets that were previously run by the Ministry became companies owning generation, transmission and/or distribution assets (with some privatization). The Ministry's regulator functions were given to the State Electricity Regulatory Commission, which reports directly to State Council. However, as some have pointed out, the power of the SERC is not necessarily in line with its reporting authority and appears subordinate to the NDRC (Andrews-Speed 2004; IEA 2006).

Ministry of Water: While coal is the dominant fuel in the electricity sector as a whole, hydropower plays an extremely important role, particularly in rural areas. As discussed elsewhere, small hydropower has been key to the electrification of the Chinese countryside. This has made the Ministry of Water an important player in the sector as it has to coordinate and approve usage of water for power and other purposes. Until the reform period, water and energy were part of the same ministry (and were again for a period of time during the reform period).(Andrews-Speed 2004; IEA 2006)

Other Ministries: A number of other central government ministries and agencies, some current and some no longer in existence, have had an influence on the Chinese electricity sector. The Ministry of Science and Technology has been heavily involved in the development of electricity generation technologies, including those for distributed generation in rural areas. The Ministry of Agriculture has had some involvement with schemes intended to reduce deforestation and provide irrigation to farmers.

Local Branches of Centralized Organizations: There are a number of central government functions that are implemented, at least in part, by regional and local branches of state organizations. At times, initiatives have actually arisen out of these sub-national branches (e.g. the IMAR Ministry of Science and Technology's role in the development of household level wind systems). At other times, they have simply acted as agents for the national organization (e.g. the role of local Ministry of Water offices in regulating small hydropower operations). However, at times they have also acted against the wishes of the central authority. Having local bureaus has opened up the system to some level of corruption and capture by local authorities.

Local Authorities: The degree to which local authorities have acted simply as an arm of the state has varied widely across time and space in China. The electricity sector is no exception. For example, local prices have often deviated from the official book rate due to the addition of various taxes and levies at the local level in order to meet local needs (IEA 2006). Similarly, attempts to shut down small coal plants were not always successful due to the reluctance of local authorities (which often owned local industries) to lose a cheap source of power, as well as a revenue stream (Wirtshafter and Shih 1990).

Key Policies

Of particular importance in the growth of rural electrification in China were specific policies put in place regarding local generation of electricity, particularly by small hydropower stations:

- "3-Self" Policy: This policy was designed to allow local entities control over their assets. It is called the "3-Self" policy because its focus was on self-construction, self-management, self-consumption. Put another way, the goal was that "who invests, and who owns, benefits."
- Rural SHP foundations: These organizations were funded by a local tax on electricity sales.
- "Electricity generates electricity"
- Tax Reduction Policy: Small hydropower owners have the benefit of both a lower value added tax and income taxes that either lowered or forfeited altogether, making their electricity cheaper and more affordable and their operations more profitable.
- Loan Policy: Significant loans have been available for small hydropower construction through the central banks. These loans have been at very low interest rates (or sometimes at zero interest) and the capital is often not paid back with little or no penalty being imposed by the bank.
- Protected Supply Areas: In theory, local producers were supposed to have protected supply areas, making them a monopoly supplier. However, as discussed elsewhere in this chapter, this did not prevent the centralized grid companies from taking over (either in part or wholly) some local operations.

(Tong 2004; Pan et al. 2006)

Small Hydropower in China

The dominant mode by which distributed electrification has contributed to the electrification of rural areas is the use of small hydropower systems (less than 25 MW). As of 2001–2002, there are approximately 42,000 small hydropower stations with a capacity of around 28 GW. Generally, discussions of the history of Chinese rural electrification divide it into three or five stages as certain aspects of the institutional environment change. However, for simplicity sake and because the data we have on small hydropower is not finely resolved enough to use all three to five time periods, we have divided the small hydropower model into two separate models. We use 1979, the time of a major change in the institutional environment as the central government devolved some powers and also began a transition to a more market based economy, as the dividing line between the two small hydropower Distributed Rural Electrification Models.

Early Small Hydropower Efforts

Prior to the reforms of 1979, which continued into the eighties, distributed rural electrification in China was primarily through development of small hydropower stations and was geared towards meeting the production needs of the state. By the end of this period, a total of nearly 7 GW of small hydropower stations had been installed.

Coordination, planning and financing came from the central government while local participation came from some local finances and labor equity (as well as land and some equity). The centralized hierarchical political and economic system also existed within the rural electrification sphere with a chain of command that passed from the central government through the provincial to lower levels of government and finally the county (Yang 2003; Barnes 2007). Ownership of these facilities, however, was vague. Particularly in the last decade of this time period, local initiatives to develop small hydropower were encouraged by the central government. In addition to developing favorable policies that promoted "self-reliance," the government provided a modest 20% subsidy and aid in technology development, thus improving the manufacturing base of the Chinese turbine industry (Tong 2004). The model was thus a hybrid one involving both the local and central governments, with a shift over time in favor of local governmental control (Tong 2004; Pan et al. 2006; Barnes 2007; CH.Int March 2005; CH.Gov May 2005).

DREM Parameters

Organization: Central and local governments
Technology: Mini-grids powered by small hydropower
Target Customers: Primarily productive activities

Financial Structure (Capital): Central government subsidized loans, direct subsidies, labor equity from local population.

Financial Structure (O&M): In principle, tariffs were supposed to be set to recover costs.

Control Variables

Capital Cost Subsidies (High): Much of the cost of small hydropower stations came from either capital grants and concessionary loans from the government and/or free equity from local labor.

Operating Cost Subsidies (None): There were no direct operating cost subsidies and, in principle, the projects should have been self-supporting.

Density of Customers (High): Small hydropower projects were designed as mini-grids serving villages and then, sometimes, later joined into larger networks of local grids.

Remoteness of Customers (High): The regions suitable for small hydropower were largely in more remote, mountainous regions.

Policy Regime (Favorable): There were numerous policies put into place to support small hydropower, including ones to improve the commercial viability of small hydropower projects and others to protect the projects from external influence.

Regulatory Regime (Favorable): The regulatory regime was continuously built upon to create effective standards and to aid in the development of small hydropower.

Outcomes

Access (High): Even before the county electrification programs of the eighties, small hydropower programs had resulted in 7 GW of power being installed.

Sufficiency (High): From early on there were efforts to include the needs of agricultural users and industrial users, meaning the power capacity levels had to be higher.

Quality (Medium): Small hydropower systems in areas of insufficient year-round water supply faced capacity constraints during dry seasons.

Sustainability (High): The strong government support and the use of the electricity for productive activities ensured that the system were maintained and kept running.

Replicability (High): While the growth in systems was slower than in the later stage, there were still large numbers of systems installed as the initial, successful, installations were replicated.

Recent Small Hydropower Efforts

The reforms of the eighties and into the nineties brought about both decentralization and a change in state-local relations as well as economic liberalization. This changed the fundamental nature of the small hydro-power distributed rural electrification model. Local authorities took over responsibility more clearly for rural electrification and small hydro plants. However, the Ministry of Water still had a large role to play in planning and approving project. At the same time, the eighties saw a rural electrification push by the central government for rural poverty alleviation and equity rather than just productivity. Beginning in 1982 with a 100 county program (later expanded to hundreds more counties), the government provided generous subsidies for meeting primary rural electrification criteria (100 billion RMB in the first phase).

Financing for small hydro projects occurred through the state-owned banks that served rural areas and they often provided low or even no-interest loans. There was also a mix of power and politics on local level. Utilities in many areas were indistinguishable from the local power structure. Where profitable, these power utilities were seen as revenue source. At the same time, like many state-owned enterprises, they had a soft budget constraint. Even when they were not making money, there was no foreclosure. In part, this seems due to the fact that the central government (whose banks had lent the money) placed a great deal of emphasis on rural electrification and would likely object. In addition, the banks found that while small hydro power companies were not necessarily safe investments, they could at least be paid the interest on the loans and were sometimes a better investment than their other rural lending options. The shareholding system put into place for ownership during the sixties was strengthened and during the latter part of this time period, private investors began putting money into small-hydro projects. A number of projects visited in Zhejiang province were privately owned and were profitable.[2] However, they were also situated in a power hungry area of the country and were connected directly to the grid system.

The mix of funding sources for small hydropower during part of this time period can be seen in the table below. Public investment remains a significant part of the picture (both in terms of direct monies from the central government and from loans), but the sources of funds is significantly more varied than the prior period in which capital was nearly all from central funds (Table 5.1).

Table 5.1 Sources of investment capital for small hydropower from 1996 to 2003. Millions of RMB (Source: China Yearbook of Electricity Industry 2004).

Year	Total	Central government	Loans	Foreign direct investment	Local government	Other (electricity levy. SHP income, private)
1996	14,489	850	6,519	809	4,222	2,089
1997	14,565	833	6,614	712	4,230	2,173
1998	15,874	879	6,907	267	5,310	2,508
1999	18,768	2,329	7,377	1,131	5,505	2,423
2000	25,078	4,375	10,400	1,378	6,722	2,201
2001	29,166	3,017	11,085	8,053	4,975	2,034
2002	24,747	4,535	11,440	355	6,132	2,282
2003	30,995	3,585	11,918	1,047	10,030	4,413
total	173,664	20,407	72,264	13,756	47,130	20,126
%	100.0	11.7	41.6	7.9	27.1	11.6

[2] At least one entrepreneur, however, was also a former local Ministry of Water official. It would be interesting to see if firms that are profitable are those that are better able to straddle both the markets and the political system, similar to the "dual firms" found in research on large independent power producers.

Theoretically this model should be sustainable since on paper it is based upon a cost-recovering tariff established by the local price bureau and based on a cost-plus accounting method. In reality many installations do not meet the cost recovery criteria. This is due to a number of factors. One important factor is the integration of these local utilities into the local government budget and lack of loan enforcement that these enterprises enjoy as a result of the priority the central government places on rural electrification and their position with the local government. Another was the take-over, without compensation, of many local distribution systems by the larger centralized grid companies (See Box 2 for an example) (Tong 2004; Pan et al. 2006; Barnes 2007; CH.Gov2 June 2005; CH.Site.SHP2 June 2005; CH.Site.SHP3 June 2005; CH.Int March 2005; CH.Gov May 2005; CH.Site.SHP May 2005).

The success of the rural small hydropower program can be seen in the number of counties that rely on the technology for at least a portion of its electricity supply. While the number of counties that rely solely on small hydro has been declining in the late nineties, there were still over 400 counties that relied solely on it for supply (some of the counties listed in other columns also have small hydropower, but it is not the primary source of electricity). Overall, estimates are that roughly 300 million people rely solely on small hydropower while up to another 400 million might rely in part on small hydropower (Tong 2004) (Table 5.2).

Box 2 One Small Hydro Power Plant in Hubei

One small hydro power plant in Hubei province illustrates both the issue of lack of loan enforcement and the struggle between local and centralized authorities over the electricity grid. Started in 1994, this relatively small plant faced two simultaneous crises. First, only a short while after being built, the grid was extended into this area and the plant was taken over by the grid company. While the generation plant was returned to the local government, the distribution system became part of the regional grid. The utility which formally was able to sell its power at 0.58 Y/kWh was now a wholesaler to the grid company at 0.17 Y/kWh. Second, due to the merger of the local government with another local government, the money for the second phase of the plant was absorbed into the new local government's coffers. The plant was still responsible, however, for the loan on the full amount. However, SHP enterprises that faced these and other problems do not fail due to strong central support for continued operation and the reluctance of banks to foreclose or seize collateral. However, it should also be noted that there is wide variability in performance (e.g. Zhejiang's profitable private plants that mainly sell to the grid vs. Hubei's smaller plants that remain locally owned) (CH.Site.SHP May 2005).

Table 5.2 Electricity supply at the county level (Source: China Yearbook of Electricity Industry 2004)

	Large grids	Wholesale	Small hydro	Small thermal
1995	707	996	567	79
1996	716	1,004	571	81
1997	727	1,005	580	66
1998	775	1,065	513	35
2000	854	1,131	433	20

DREM Parameters

Organization: Local Governments or Private
Technology: Mini-grids powered by small hydropower
Target Customers: Productive activities plus households
Financial Structure (Capital): Mix of funds, including loans and equity
Financial Structure (O&M): Tariffs set to cover costs.

Control Variables

Capital Cost Subsidies (Medium): While central government funds were significant in this period of time, there has been a steady rise in the use of private capital as the system moved to a more market-oriented structure.
Operating Cost Subsidies (None): As in the earlier period, small hydropower projects were expected to have cost-recovering tariffs.
Density of Customers (High): The small hydropower systems are based on micro-grids which require a certain customer density to be viable.
Remoteness of Customers (High): Small hydropower is continuing to be used to provide electricity in remote mountain regions (as well as some areas closer to industrial centers).
Policy Regime (Favorable): Earlier policies that were favorable to small hydropower were continued and new policies were also put in place to rapidly expand capacity.
Regulatory Regime (Favorable): The regulatory regime continued to favor the development of hydro resources.

Outcomes

Access (High): The county electrification programs and other efforts (including private sector developments) have resulted in high rates of electricity access.

Small hydropower is the primary source of electricity for hundreds of millions of rural Chinese.

Sufficiency (High): Power is provided for both household use as well as for commercial and industrial activities.

Quality (High): Continued interconnection between small hydropower systems and connection to the main grid has helped deal with the water availability problem.

Sustainability (High): While some small hydropower systems do not exhibit the cost recovery characteristics expected of them, the likelihood of foreclosures is effectively zero.

Replicability (High): The model has shown itself to be highly replicable within China and many aspects of the model should be replicable elsewhere. However, some of its success is certainly due to the financial capacity of the central state to provide loans and the particular structure of the banking system in China, which resulted in low or no interest loans and little penalties for lack of repayment.

The Township Electrification Program

By contrast with both the pre- and post-1979 electrification successes using small hydro power, the Township Electrification Program illustrates clearly the problem with some top–down models. The Township Electrification Program (*Song Dian Dao Xiang*) is part of China's larger rural electrification effort, the Brightness Program. TEP set and achieved the ambitious target of electrifying over 1,000 townships (roughly 1 million people) in just 3 years using a combination of small hydropower (378 villages, 200 MW), PV and some hybrid systems (666 PV villages and 17 hybrid villages totaling 20 MW). It was able to do so, in part, because it harnessed the power of the central government. This was a program of the National Development and Reform Commission, executed through contracts with private and semi-private firms. All capital expenses were paid for by the central government and contractors were obligated to provide 3 years of service. Coverage of O&M during those 3 years was, at least in part, recovered by tariffs, but those tariffs had no uniformity and varied widely (including some customers paying no fees for the first 3 years).[3] No provisions were made in the program for coverage after 3 years. Furthermore, ownership of the

---

[3] In the systems installed by one of the contracted companies, user fees ranged from Yuan 0.4 to 1/kWh ($.05–0.12/kWh). These revenues were collected by the townships from 99% of households, but government buildings and schools were not charged. The low consumption of rural users also meant that even though a tariff was charged, revenue was low. CH.Ind.Ren (May, 2005). Author Interview with Renewable Electricity Entrepreneur. Beijing, China.

systems remains an open question. The target customers were households even though entire townships were being electrified. However, no mechanism was put in place to transfer the systems to local ownership after the initial contract terms expired. The program was highly successful in providing short term electricity access to 1,061 townships (exceeding original project goals). However, the lack of clear ownership and cost recovery mechanisms jeopardize the sustainability of these systems over time. The emphasis of the Chinese government on rural electrification and the high visibility of this program make it unlikely that systems would be allowed to fail in the short term, but over the longer term new institutional arrangements will have to be made. It would also be difficult to replicate this model as it relies entirely on the central government for financial support and coordination. This low replicability is reflected in the fact that the Chinese government would like to conduct a second phase to electrify smaller programs but has commissioned studies to determine how the model should be changed to avoid these problems (Ku et al. 2003; Energy Research Institute 2005; Ch.Int.Don2 March 2005; Ch.Int.Don March 2005; CH.Ind. Assc March 2005; CH.Ind.Ren May 2005).

DREM Parameters

Organization: The National Reform and Development Commission through con-tracted private firms.
Technology: Mini-grids based on a mix of renewable technologies (though pri-marily small hydropower).
Target Customers: Households + Community buildings
Financial Structure (Capital): Central government funds
Financial Structure (O&M): Unaccounted for in planning, mix of no to com-plete cost recovering tariffs by the contractors during the initial 3 year phase.

Control Variables

Capital Cost Subsidies (High): Capital costs were provided by the central state.
Operating Cost Subsidies (High): Operating costs for the first 3 years were gen-erally provided as part of the initial contract, though there is some variation between projects.
Density of Customers (Medium): The target communities were townships, which are the fourth level of administration in the Chinese system and contain within them villages and communities.
Remoteness of Customers (High): The target communities were the remaining townships that had not been fully electrified in previous programs and tended to be more remote.
Policy Regime (Favorable): This was a program of the highest level authorities in China, the National Development and Reform Commission.

Regulatory Regime (Favorable): Small hydropower was a key component of the program and benefited from the favorable regulations discussed above. There were no regulatory barriers to implementation of micro-grids based on wind or solar either.

Outcomes

Access (Medium): While electrifying 1,000 townships may seem like a significant change in access, these electrified populations within these townships represent a small fraction of the rural population, most of which already has access. This project was large, but it was mainly a way to try to capture the last remaining pockets of poor electrification.

Sustainability (Medium): The model would likely not be sustainable under any other context given the lack of a pre-planned organizational structure after the initial phase of the projects and lack of tariffs in many cases. However, as an important project of the Chinese government, it is likely that solutions will be found to these problems to ensure that the projects continue to operate.

Replicability (Low): The model was highly dependent upon funds from the central government. There was no contribution from consumers or the private sector for construction and in many cases no tariff payments for covering operational costs.

The Market for Small Solar Home Systems

China has one of the largest cash markets for solar home systems in the world with an estimated 150,000 units being sold in 2001 (Drillisch et al. 2005). These units are usually rather small in power (in the range of 20 W). This likely reflects similar dynamics as the cash market in Kenya where households cannot afford larger systems as a single purchase. Some may purchase the system in a modular way, building up their capacity as income allows. A significant shift in this model occurred in 2001 with the start of a joint project of the Chinese Government's National Development and Reform Commission, the Global Environmental Facility and the World Bank. A significant portion of the China Renewable Energy Development Project (REDP) was geared towards improving certain aspects of the solar home system market in China. This was done through a variety of means, but we will focus here on two. First, support was given to various manufacturers in the industry for improving the quality of their product in a certified manner as well as other technical and institutional support for the industry. Second, a subsidy was given to the manufacturers for sales that met certain eligibility criteria. The subsidy was initially $1.5/W and then raised to $2/W (REDP 2005; REDP 2006). These activities did not fundamentally alter most aspects of the Distributed Rural Electrification Model other than the financial structure through the subsidy and the technological fit, by improving product quality. Since most elements of the model remain the same and the subsidy is going to be phased out, we have included both the pre-REDP cash market and the REDP in this description, being careful where necessary to distinguish between the subsidized and unsubsidized cash market.

This model is based primarily on local retail dealers selling the products of national or regional level photovoltaic equipment companies. The dealers may have a multi-purpose shop in which various products are sold, PV modules just being one. However, a significant fraction of the systems sold have been through dealers and manufacturers that have access to the REDP program (accounting for roughly 50,000 units per year). This model thus represents a hybrid model which can be best characterized as centralized and primarily non-governmental but with strong central government support in its later phase. It is considered centralized because the local dealers have little to do with the decision making regarding pricing, technology, and implementation. For them, solar panels are just another product to be sold.

Financially, this model is very sustainable as all costs are covered under the cash sale. Even the REDP subsidy should not make much difference. In 2001 the SHS unit price was \$12–13/Wp. This means the subsidy amounted to 15% or less of the capital costs. However, the improvements in manufacturing and market competition have actually resulted in higher quality modules being sold at a lower price than before the program, roughly \$7.5–8.5/Wp (REDP 2006). Given that consumers were purchasing lower quality components at a higher price than this before the subsidy, there is no reason to believe that this market cannot continue to sell SHS once the subsidy is removed. It should also be noted that the improvements made to upstream suppliers has also allowed them to improve the sales outside of the REDP SHS market, improving their financial viability and further increasing the probability of sustaining this market. This model has already demonstrated its replicability with numerous manufacturers and independent dealers involved (even prior to the REDP program).

Additional Sources: (Ch.Int.Don2 March 2005; Ch.Int.Don March 2005; CH. Ind.Assc March 2005)

DREM Parameters

Organization: Commercial sales through retail outlets
Technology: Solar home systems
Target Customers: Households
Financial Structure (Capital): Cash sales including a modest subsidy
Financial Structure (O&M): Consumers

Control Variables

Capital Cost Subsidies (Low): Subsidies were modest and tied to performance criteria.

Operating Cost Subsidies (None): No operating subsidies were given. Customers are responsible for operating costs (though there was an initial warranty period).

Density of Customers (Low): The programs sold individual solar home system units to customers that were unlikely to be able to access a local grid.

Remoteness of Customers (High): The customers were located in remote regions, particularly in the western region of China.

Policy Regime (Favorable): The program to expand solar home system sales in China was supported by and partially implemented by the Chinese government.

Regulatory Regime (Favorable): The rules governing quality standards under the program acted effectively as a regulation that favored participants in the program as they were able to access the subsidy.

Outcomes

Access (High): The Chinese solar home system market is one of the largest in the world, with annual sales reaching as high as 80,000 units per year during the time period of the program.

Sufficiency (Medium): Systems are of a smaller size, making them more affordable but only able to provide power for limited applications. Given the economic status of most customers and the ability to purchase additional units for more power, the systems should be able to meet most needs.

Sustainability (High): Customers have made an initial investment, which they will want to maintain through the necessary purchase of replacement equipment. Systems are in place for maintenance and spare parts purchasing.

Replicability (High): The limited level of the subsidy and its mandate for encouraging competition and improving quality means that the resources required were not excessive given the outcomes. Overall, the model should be replicable. However, the full effect would only be seen in contexts where multiple suppliers of solar home systems could exist and be sustained.

Household Wind Power in the Inner Mongolian Autonomous Region

Another cash market for household based renewable energy exists in the Inner Mongolia Autonomous Region where wind conditions are favorable. Beginning in 1980 the local branch of the Ministry of Science and Technology began a program to develop 50 W and 100 W household scale wind turbines for rural Mongolian herders. Support was given for research and technology development, as well as a program of technology transfer from abroad. The herders are very dispersed but have higher incomes than many other rural Chinese due to the high value of their commodities, particularly items like cashmere. In 1986 the technology development program was supplemented with a 10% subsidy towards the capital costs of a wind turbine. A voucher was obtained from the local energy office and taken to the manufacturer when purchasing the unit.

The results of this technology development and small subsidy program have been impressive. By 2004, some 140,000 units had been sold and up to 80% of households in some areas were using household wind generators. This is out of a total of 440,000 households in rural areas. Direct sales by the rural energy companies are on the order of 4,000–5,000 per year. However, this power was not cheap with energy cost of $0.55/kWh according to Lew and $0.25–0.37/kWh according to Zhou et al. (Lew 2000; Zhou and Byrne 2002; Stroup 2005).

While the wind program in IMAR has proven to be quite successful, there were some problems. First, the wind-only systems suffered during the summer when winds were low. Second, the car batteries used with these systems were not designed for the charging and discharging cycles to which they were subjected.

DREM Parameters

Organization: Hybrid between private enterprise and the regional government
Technology: Household scale wind turbines
Target Customers: Households
Financial Structure (Capital): Cash market with small subsidy
Financial Structure (O&M): Consumers

Control Variables

Capital Cost Subsidies (Low): Capital subsidies existed but were only a modest portion of overall capital costs.
Operating Cost Subsidies (None): No operating subsidies were provided.
Density of Customers (Low): The large nomadic population of IMAR is extremely spread out.
Remoteness of Customers (High): The population of IMAR is remote.
Policy Regime (Favorable): Support for the program came from both the IMAR and the central government.
Regulatory Regime (Favorable): Support for the program came from both the IMAR and the central government. Furthermore, there were no regulations that posed problems for implementation.

Outcomes

Access (High): There was a very high rate of uptake among the herders, with well over a majority of households having these systems in some areas.

Sufficiency (Medium): The system capacity was sufficient for most household needs, but dependent upon availability of wind.

Quality (Medium): The dependence on intermittent (and seasonally variable) windpower leads to some quality issues in terms of availability

Sustainability (High): There are no indications of widespread problems with the systems. The households that purchase these systems can also afford the regular battery replacement costs and other ongoing expenses.

Replicability (High): There is clear evidence based on its success, that the model was replicable throughout IMAR. The low subsidy means that replicating the model elsewhere would not have to rely upon large public resources. However, it would only work in areas with a relatively wealthier rural population and with good wind resources.

Household Wind/Photovoltaic Hybrid Systems in the Inner Mongolian Autonomous Region

As a result of the success of the wind power program in the Inner Mongolian Autonomous Region, a new Wind/PV hybrid program was initiated by the US DOE (NREL), MOST, the Chinese Academy of Sciences and the Center for Energy and Environmental Policy at the University of Delaware.

A pilot project installed 400 between 1996 and 2001, with another 8,000 installed afterwards. The system size was large (400–500 W), allowing it to meet household needs, including refrigeration in the summer. However, it was also expensive, even for this area. The families selected for the pilot were close to installation and maintenance services. The subsidy was higher than in the original wind program, 30% off a Y10,000 system. The local MOST offices were the installers and maintainers of the system. However, there has been a concern about the ability to maintain the technology over time and the financial structure of the program (given the higher subsidy rate) (Ch.Int.NGO March 2005).

In 2004 a small sample survey was conducted to determine the impact of the program. The survey found 50% of the units in good condition, 35% had experienced periodic failures and 16% were in poor condition. One significant problem was with poor quality inverters. There were also complaints about short battery life and having to replace batteries. All units had gone through one battery replacement. While there were complaints about cost, these costs were relatively low compared to incomes and costs of other major appliances. However, there was a major difference in willingness to pay for replacement parts rather than an appliance. The estimated energy costs range from $0.30/kWh to $0.67/kWh (Lew 2000; Zhou and Byrne 2002; Stroup 2005).

DREM Parameters

Organization: Regional government plus international donors
Technology: Hybrid of household scale wind and photovoltaic systems

Target Customers: Households
Financial Structure (Capital): Cash market with modest subsidy
Financial Structure (O&M): Consumers

Control Variables

Capital Cost Subsidies (Low): Capital subsidies existed but were only a modest portion of overall capital costs.
Operating Cost Subsidies (None): No operating subsidies were provided.
Density of Customers (Low): The large nomadic population of IMAR is extremely spread out.
Remoteness of Customers (High): The population of IMAR is remote.
Policy Regime (Favorable): Support for the program came from both the IMAR and the central government.
Regulatory Regime (Favorable): Support for the program came from both the IMAR and the central government. Furthermore, there were no regulations that posed problems for implementation.

Outcomes

Access (Medium): The higher costs of these systems limited their affordability, even among the richer IMAR herders.
Sufficiency (High): The capacity of these systems allowed household electricity needs to be met.
Quality (High): The use of two complementary technologies improved quality. There were, however, some complaints regarding components.
Sustainability (High): The vast majority of systems were considered to be in good shape or only had experience periodic failures (to be expected under the circumstances).
Replicability (High): The replicability is high within the context of IMAR and China, given the support of the government and the financial resources of the target population. Outside of China, there may be an issue as the technology is relatively expensive and IMAR represents a best case scenario (good and complementary resources and higher income population).

Other Models

A small number of biomass gasification for power generation have been installed. However, there is no information on who installed these 150 units, where and for what purpose. Capital costs are low ($320–430/kW) and O&M costs were estimated at up to 17 c/kWh, which would probably make them cost competitive with some alternatives (Leung, Yin et al. 2004). Another DREM for which it is difficult to get information is independent diesel generating mini-grids and battery chargers. Batteries and diesel generators apparently provide power for 77 million people in 30,000 villages, as of 2003 (The Institute for Development Studies 2003). However, specific information has not been found to outline the parameters of the business model and their performance.

Conclusions

China's successes in distributed rural electrification provide a number of lessons that are useful for this study. Chinese models that emphasize cost recovery, such as the renewables markets for wind, PV and hybrids shows high levels of access, sustainability and replicability despite the fact that these systems can be expensive. In the case of the wind and hybrid systems in Inner Mongolia, the ability to pay problem is lessened by the profitable nature of the rural economy. For the PV systems, a market based on small size modular units allows systems to be affordable.

In principle, the small hydro systems, which have had the greatest success in increasing access, should be profitable and be considered along with the wind and PV markets. These systems provide integrated electricity services that include productive activities (and were primarily installed to meet productive needs originally) and the central government support in terms of subsidies and low interest loans reduced their costs. These systems contributed greatly to the growth of agriculture and later industry in rural areas and were supposed to be able to charge cost-recovering tariffs. However, many systems lose money due to local factors as well as a shifting policy environment and poor property rights. Despite the fact that many systems are not profitable and do not recover their costs, the strong institutional backstop provided by the central government allows them to continue to operate.

It is clear that the Chinese government has played a large role in the success of the rural electrification efforts across China. However, that role has varied from DREM to DREM, and in some cases it has been positive and in others negative. When the central government has played a secondary supporting role through policies and regulations, the results have been to increase access and generally to increase sustainability and replicability. Its support for technology development in small hydropower, photovoltaics and small wind turbines has led to cost decreases and improved quality. Similarly, its provision of capital subsidies, creation of protective policies, and low interest loans (through the central banks) helped the growth of the small hydro systems. On the other hand, when the Chinese government has involved itself more directly, the outcomes have been less favorable for distributed rural electrification efforts. This can be seen most clearly in the Township Electrification Program, a completely centralized top–down effort by the government which provided all of the funds and made no provision for the long-term management of the systems.

This also highlights the important role played by local organizations. In the case of small hydropower, it was originally the county branches of the central government, but later it was local governments and their utilities that were the primary owners of the systems, as well as some private owners. The Inner Mongolia success is also based, in part, on local offices of the government along with local manufacturers and dealers.

Clearly within China this model of primarily local ownership of small hydro mini-grids has been a highly replicable DREM with tens of thousands of installations. Outside of China it may be difficult for two reasons. First, it relied on the strong central resources of the state through directed policies and occurring in a context in which local governments are extensions of the central government so coordination problems may be minimized (though when policies conflict, as they did in 1999, the results can be disastrous). During the process of decentralization, the central government was able to shift its involvement to a different support role. Second, the problems that occurred due to inconsistencies in policies, resulted in systems that were not necessarily recovering their costs. They maintained sustainability due to the strong central government. In essence, this muddling through of problems that would sink programs in other countries is a feature of the Chinese institutional environment that would be difficult to replicate elsewhere.

One factor that could be replicated elsewhere was China's early emphasis on rural electrification as an integral component of its agricultural modernization and later industrialization policy. The early small hydropower projects were intended primarily for enhanced productivity and rural development, with a focus more explicitly on household electrification coming much later. By tying electricity development to rural development, the Chinese government appears to have been able to have success in both. Rural agriculture and industry was able to get the electricity it needed to be more productive (or function at all in those settings) and, in turn, these customers provided a strong base for the electrification effort (i.e. increased demand to justify investment and the possibility of revenues for electricity generation). Of course, in reality, the linkage has not been as smooth as it could have been and it is only because of the strength of the central system and a lack of accountability at some levels that some of the facilities have continued to operate.

It also instructive that a significant portion of China's rural electrification has been through various types of renewable energy technology ranging from the early hydropower plants to the more recent solar home systems and hybrid systems. However, in all cases, the emphasis was not on renewable energy as an environmentally beneficial technology choice but rather on technologies that were well suited to local conditions. The result has been highly successful dissemination of renewable energy into the rural areas of China, a feat that has been hard to replicate elsewhere. It is often forgotten that China is the world leader in installed renewable energy capacity (excluding large hydropower) as a result of its small hydro program (REN21 2008) (Table 5.3 and 5.4).

Table 5.3 Summary table for Chinese distributed rural electrification models

DREM parameters		SHP-early	SHP-recent	Wind power – Inner Mongolia
DREM parameters	Organization	Central-local government hybrid	Local-central government hybrid + private	Private-regional government hybrid
	Target customers	Primarily productive	Productive plus households	Households
	Technology	Small hydro power mini-grids	Small Hydro Power mini-grids	Wind
	Financial: capital	Central gov. subsidies/loans plus labor equity	Mix of funds	Consumers with small subsidy
	Financial: O&M	Supposed to be cost-recovering tariff	Supposed to be cost-recovering tariff	Consumers
Control variables	Capital cost subsidieS	High	Medium	Low
	Operating cost subsidies	None	None	None
	Density of customers	High	High	Low
	Remoteness of customers	High	High	High
	Policy regime	Favorable	Favorable	Favorable
	Regulatory regime	Favorable	Favorable	Favorable
Outcomes	Access	High	High	High
	Sufficiency	High	High	Medium
	Quality	Medium	High	Medium
	Sustainability	High	High	High
	Replicability	High	High	High
Notes on institutional factors	Policy measures	Numerous favorable policies	Numerous favorable policies	Subsidy and industry support policy
			1998 policy decision for centralized take-over of grids	
			Rural grid renovation policy to improve quality	
	Regulatory Measures	Favorable regulations	Favorable regulations	
	Other	Relationship with local government	Change in state-local relations	Higher income consumers

Table 5.4 summary table for Chinese distributed rural electrification models (continued)

DREM parameters		Wind/PV hybrids in IMAR	TEP	PV cash
	Organization	Regional government and international	Central government	Private central-local hybrid
	Target customers	Households	Households + Community buildings	Households
	Technology	Wind/PV hybrid	Renewable mini-grids	PV Solar home systems
	Financial: capital	Consumers with modest subsidy	Central government	Cash Sales plus modest subsidy
	Financial: O&M	Consumers	Unaccounted for in planning	Consumers
Control variables	Capital cost subsidies	Low	High	Low
	Operating cost subsidies	None	High	None
	Density of customers	Low	Medium	Low
	Remoteness of customers	High	High	High
	Policy regime	Favorable	Favorable	Favorable
	Regulatory regime	Favorable	Favorable	Favorable
Outcomes	Access	Medium	Medium	High
	Sufficiency	High		High
	Quality	High		High
	Sustainability	High	Medium	
	Replicability	High	Low	
Notes on institutional factors	Policy measures	Subsidy and industry support policy	Project of central government	Subsidy and industry support policy
	Regulatory Measures			
	Other	Higher income consumers	Sustainability due to government priority on rural electrification.	

References

Andrews-Speed P (2004) Energy policy and regulation in the People's Republic of China. Kluwer Law International, The Hague

Barnes DF (ed) (2007) The challenge of rural electrification: strategies for developing countries. Resources for the Future, Washington, DC

CH.Gov (May 2005). Author interview with Ministry of Water official. Beijing, China.

CH.Gov2 (June 2005). Author interview with county officials. Zhejiang, China.

CH.Ind.Assc (March 2005). Author interview with Renewable Energy Industrial Association. Beijing, China.

CH.Ind.Ren (May 2005). Author interview with Renewable Electricity Entrepreneur. Beijing, China.

CH.Int (March 2005). Author interview with International Industry Centre. Hangzhou, China.

Ch.Int.Don (March 2005). Author interview with international donor. Beijing, China.

Ch.Int.Don2 (March 2005). Author interview with international donor. Beijing, China.

Ch.Int.NGO (March 2005). Author interview with international non-governmental organization. Beijing, China.

CH.Site.SHP (May 2005). Author site visit to small hydropower project. Hubei, China.

CH.Site.SHP2 (June 2005). Author site visit to small hydro project. Zhejiang, China.

CH.Site.SHP3 (June 2005). Author site visit to small hydro project. Zhejiang, China.

China Yearbook of Electricity Industry (2004) China yearbook of electricity industry, 1993–2003. Electricity Industry Press, Beijing

Drillisch J, Dubois a et al (2005). Energy policy framework for electricity markets and renewable energies in the PR China. Eschborn/Beijing, TERNA Wind Energy Programme, Division Environment and Infrastructure, Deutsche Gesellschaft für Technische Zusammenarbeit (GTZ) GmbH, p 25

Energy Research Institute (2005) UNDP supported baseline survey for the Song Dian Dao Xiang Program (the Township Electrification Program of China). Energy Research Institute, National Development and Reform Commission of the People's Republic of China, Beijing, China, p 69

IEA (2006) China's power sector reforms: where to next? Organization for Cooperation and Development (OECD), International Energy Agency (IEA), Paris

IEA (2009) The Electricity Access Database. International Energy Agency. http://www.iea.org/weo/database-electricity/electricity-access-database.html

Ku J, Lew D et al (2003) Sending electricity to townships: China's large-scale renewables programme brings power to a million people. Renewable Energy World, pp 56–67

Leung DYC, Yin XL et al (2004) A review on the development and commercialization of biomass gasification technologies in China. Renewable Sustainable Energy Rev 8:565–580

Lew JD (2000) Alternatives to coal and candles: wind power in China. Energy Policy 28(4):271

Lieberthal K, Oksenberg M (1986) Bureaucratic politics and Chinese energy development. U.S. Dept. of Commerce, International Trade Administration, Washington, DC

Oi JC (1999) Rural china takes off: institutional foundations of economic reform. University of California Press, Berkeley, CA

Pan J, Peng W et al (2006) Rural electrification in China 1950–2005. Research Centre for Sustainable Development, Chinese Academy of Social Sciences and Program on Energy and Sustainable Development, Stanford University, Beijing, China and Stanford, CA

REDP (2005) REDP Progress Report (2001–2004). NDRC/GEF/WB China Renewable Energy Development Project, Project Management Office, Beijing, p 21

REDP (2006) NDRC/GEF/The World Bank: China renewable energy development project. NDRC/GEF/The World Bank China Renewable Energy Development Project, Beijing, p 35

REN21 (2008) Renewables 2007 Global Status Report. REN21 Secretariat and Worldwatch Institute, Paris and Washington, DC, p 51

Rufin C, Rangan US et al (2003) The changing role of the state in the electricity industry in Brazil, China, and India: differences and explanations. Am J Econ Sociol 62(4):649–675

Steinfeld ES (1998) Forging reform in China: the fate of state-owned industry. Cambridge University Press, Cambridge

Stroup KK (2005) DOE/NREL Inner Mongolia PV/wind hybrid systems pilot project: a post-installation assessment. National Renewable Energy Laboratory, Golden, CO, p 106

The Institute for Development Studies (2003) Energy, poverty, and gender: a review of the evidence and case studies in rural China. The Institute for Development Studies, The University of Sussex, U.K. for the World Bank, Sussex, p 159, + Appendices

Tong J (2004) Small hydro power: China's practice. China WaterPower Press, Beijing

Wirtshafter RM, Shih E (1990) Decentralization of China's electricity sector: is small beautiful? World Dev 18(4):505–512

Yang M (2003) China's rural electrification and poverty reduction. Energy Policy 31:283–295

Zhou A, Byrne J (2002) Renewable energy from rural sustainability: lessons from China. Bull Sci Tech Soc 22(2):123–131

Chapter 6
Understanding Success and Failure in Distributed Electrification

Keywords Case study conclusions • Centralization/decentralization • Distributed electrification models • Subsidies

Introduction

The previous three chapters provided detailed information on the performance of the business models within each of the three countries included in the study. Each country had unique institutional arrangements and different types of business models, leading to quite different outcomes. However, this variation can also be used to determine whether there are overall trends in the performance of these different business models. This chapter uses the results from case studies in all three countries to draw conclusions regarding success and failure in distributed rural electrification. In the next three sections, the initial hypotheses developed in Chap. 2 are examined to determine whether they are supported by the evidence. In order to do this, we examine the impact of the business models on changes in electricity service, sustainability and replicability.

Changes in Electricity Service

Each of the four independent variables that make up the business models was hypothesized to have an effect on the changes in electricity service that result from implementation of the model. The main hypotheses are summarized in Table 6.1 (below).

The evidence from cases in all three countries supports these hypotheses with the exception of the hypothesis on Technology Choice. Certainly the widespread use of diesel generator based mini-grids in Cambodia (with secondary evidence for their wide use in Brazil and China) indicates that minigrids can be widely implemented, even at the local level by entrepreneurs or other local actors. At the same time, the success of the Chinese household renewable energy markets for solar and wind systems shows that individual installations can also contribute to broader electrification.

As expected, large gains in rural electrification over the short term were largely the result of programs by centralized organizations that had the reach and the

Table 6.1 Summary of hypotheses for electricity service

Organizational form	Technology choice	Target customers	Financial structure
Centralized, particularly governmental, programs will result in the greatest *short-term* increase in electricity access for rural households	Programs for individual installations (e.g. solar home systems) will result in the greatest *short-term* increase in electricity access	Programs that include electrification of productive activities will have higher levels of access and service quality	Programs with high subsidies will have the highest short-term impact on access

resources to target multiple areas within a country and had significant resources at their disposal.

The results are mixed when it comes to the role of Target Customers. In some cases, including productive activities benefited in creating a viable model that could improve access. This is true of the Cambodian electricity entrepreneurs that rely on segmenting the market to make their grid systems viable. It is also true of the Chinese mini-grid systems run primarily by local governments that built upon their serving local agriculture and industry. However, it is not true of the large rural electrification programs run by the Brazilian utilities or programs such as the Township Electrification Program in China. These programs are able to serve wide swaths of the population without particularly taking into account productive activities (and, indeed, are focused less on those customers and more on households).

Sustainability

All of the hypotheses regarding sustainability were broadly true, though with varying degrees of support (Table 6.2). As evidenced by the success of decentralized models in all three countries, local actors can take into account the particular needs of users within a community or can target specific users in order to be sustainable. BRASUS and IDEAAS in Brazil target productive and higher income users respectively. Similarly, in Cambodia, the REEs include productive users and not just households while Khmer Solar targets households at the higher end of the income scale.

With regard to the choice of technology, there was no evidence that fuel price variability resulted in higher sustainability for renewable technologies over fossil fuel based systems. The widespread use of unsubsidized diesel in Cambodia (and also in Brazil) shows that while high fuel prices might increase the cost of electricity, it is not sufficient to deter users and its variability had not resulted in a switch.

The hypothesis regarding maintenance of renewables did have some support. The maintenance requirements of renewable systems continue to be underestimated in some quarters, most easily evidenced by the Japanese PV/Hydro project in Cambodia. However, the success of the Chinese systems shows that widescale

Table 6.2 Summary of Hypotheses for Sustainability

Organizational Form	Technology Choice	Target Customers	Financial Structure
Decentralized organizations will be more sustainable than centralized models, particularly when accounting for institutional factors important for sustainability	Renewable technologies have the potential to be more financially sustainable due to being immune to fuel price variations Non-renewable technologies are more sustainable institutionally due to the unfamiliarity of renewable technologies and lack of associated widespread maintenance mechanisms	Sustainability (both financial and institutional) will be higher when productive activities are included	Models with full cost-recovery (particularly of operating expenses) will be the most sustainable financially

diffusion of renewables is possible and that they can be maintained if proper care is taken to incorporate it into the model.

As expected, serving productive consumers (either alone or as part of a communal electrification scheme) increases sustainability. This appears to be due both to the higher ability to pay of productive consumers and the more consistent power needs of productive actors. It is this combination that allows the REEs in Cambodia to have a viable business. Some organizations, like BRASUS in Brazil, focus exclusively on consumers that can pay back their loans through income generating activities. Another factor may be that the highly centralized efforts to expand access (e.g. in Brazil) do not tend to pay attention to small-scale productive activities in an effort to reach more households. Thus, productive consumers have to find alternatives and this may, in fact, provide more buy-in to make sure it is sustainable, since they know they do not have the government program to fall back on necessarily.

Finally, as expected, models that do not recover costs, particularly operations and maintenance costs either fail or depend upon centralized subsidies. Failures include the Japanese PV/Hydro project in Cambodia and the solar battery charging project in Brazil. Models that continue to survive solely because of subsidies (direct or indirect) include all of the utility based models in Brazil and the Township Electrification Program projects in China.

Replicability

The results provide mixed support for the replicability hypotheses (Table 6.3). Replicability is certainly possible with more centralized models, contradicting the original hypothesis. Brazil shows evidence of both replicable and non-replicable centralized models. However, it does require long-term support, including subsidies, which is difficult (though not impossible).

Table 6.3 Summary of Hypotheses for Replicability

Organizational Form	Technology Choice	Target Customers	Financial Structure
Decentralized models will be more replicable than centralized models	Individual installations will be more replicable than mini-grids. Non-renewable mini-grids will be more replicable than renewable based mini-grids	Models that emphasize individual users (productive users or households) will be more replicable than those that emphasize groups of users. Models that emphasize productive activities will be more replicable than those that only focus on households	Models that include full recovery of capital expenses will be the most easily replicable, particularly when the model includes some method for financing capital, but are limited by the willingness and ability to pay of consumers. Models with modest and targeted subsidies will be the most replicable overall

The technology choice hypothesis could not be supported as there was ample evidence of replicability for individual and mini-grid based models and for renewable and non-renewable models. Evidence could be found across cases in all three countries, ranging from a number of solar home system programs (Brazil, Cambodia and China) to renewable based mini-grids (e.g. small hydropower in China).

Models in which productive activities were the target customers (or at least included in a mix of customers) do show greater replicability for much the same reasons as including them makes a model more sustainable. These are customers that have higher incomes and have demonstrable gains from using electricity. Some models, such as the BRASUS one in Brazil specifically target productive actors. Other successful models that include productive actors are the village based utilities in China and the rural electricity entrepreneurs in Cambodia. There were no models in this study that included productive actors and were unsuccessful and not replicable (as compared to models that targeted only households or only community structures).

The results show strong support for the hypotheses regarding financial structure of the business model. Capital cost recovery is a key component in replicability. It would also seem that both willingness and ability to pay are largely underestimated, as evidenced by the results from Cambodia as well as models in other countries in which full cost recovery is achieved despite high tariffs and prices. However, subsidies do appear to help and so modest subsidies that aren't a large burden can be expected to increase replication (e.g. similar to those used to spur the growth of the solar home system in western China).

Key Findings and Conclusions

Brazil's distributed rural electrification experience has been dominated by highly centralized efforts. This includes government ministry programs for electrifying community buildings and utility run programs for household electrification. The utilities in Brazil are required to serve all customers in their monopoly service territory, whether through grid extension or through the installation of distributed systems and charge rates that are below cost of service. The historically slow pace of rural electrification, however, has led to a number of models outside the utility and government systems, including solar home system leasing schemes, cooperatives, private diesel micro-grids and renewable energy based systems for agricultural producers. The recent push for universal electrification by the government (through the utilities), which has included high capital costs subsidies for the utilities, does pose a threat to these other models.

The Cambodian situation differs remarkably from the Brazilian one. The Cambodian government is not heavily involved as of yet in rural electrification. Official electrification rates remain low (15%) and the state utility only serves the largest population centers. Despite a low official rate of electrification, the number of households with access to at least a minimal amount of electricity (e.g. enough to run a lightbulb and maybe a small television) is extremely high (50% of the households have a television and an estimated 85–90% have a lightbulb). Their electricity primarily comes from rural electricity entrepreneurs that run diesel based micro-grids, battery charging stations or a combination of the two. Prices are high to cover costs, but consumption is low and overall monthly expenditures are kept low. Some of the entrepreneurs are licensed by the government and included in the official 15% statistic, but most are unlicensed operators. Cambodia also has a small solar home system market that serves slightly wealthier consumers and has had some donor projects.

China has had stunning success in rural electrification since the early eighties. From a starting point of hundreds of millions of people without electricity, the electrification rate is now nearly 100%. Small hydropower has played a major role and currently provides power for over 300 million rural Chinese (more if one includes systems with only partial hydropower supply). The government was heavily involved, but often in a supporting role through low interest loans and guarantees and technology programs. Technology development and support has also been key to China's success in creating rural markets for solar home systems and small wind and wind/pv hybrid systems. Recently the Chinese government instituted the Township Electrification Program, which was a much more top-down centralized effort. While successful in the short term, the future of the installed systems is in doubt due to lack of local involvement and planning for post-installation operations.

This study of distributed rural electrification in Brazil, Cambodia and China shows that a wide diversity of distributed rural electrification models exist and can be successful in providing electricity to remote rural populations. However, while their individual successes and failures are not always due to the same set of factors, there are some common themes and clear trends that emerge.[1] The various models examined in this study can be roughly divided into four broad categories:

Government/Donor Model: These models, such as the Chinese Township Electrification Program or the Brazilian Programa de Desenvolvimento Energético de Estados e Municipios (PRODEEM), are directed and funded centrally. They generally aim to provide the bare minimum electricity for households or community structures. These models are characterized by high subsidies for capital costs and low tariffs. Not only do these tariffs not cover capital costs, they are often insufficient to cover operating costs. They are only able to be sustainable when the operating subsidies are sufficient. The result is high increases in access, but mixed results on sustainability and replicability. Those that are sustainable and replicable are heavily dependent on ongoing financial support. In effect, there are two sub-models, a "**Technology Dump**" model in which large-scale diffusion programs are not supported with sustainable institutions and/or subsidies and a "**Sustainably Centralized**" model in which governments or donors create viable on-going mechanisms for sustainability.

Utility Model: This model is observed in Brazil, where centralized utilities are implementing distributed generation technologies to meet their regulatory obligations. As with the government/donor model, the utility model is dependent on high capital subsidies and focused on meeting basic needs in rural areas. Unlike the government/donor model, operating costs are generally covered through a combination of tariffs and cross-subsidies. However, since the customer base for rural distributed generation is lower income, the tariffs are kept artificially low, making the utility model also dependent upon the subsidies to be sustainable and replicable.

Mixed Model: In the mixed model, implemented particularly in China, the government acts as a strong supporter of distributed generation technologies without organizing and implementing programs centrally. China's support for its renewables industry, such as the PV market and the use of household wind power in Inner Mongolia, are prime examples of this model. The government provides incentives and technical support for technology development. This creates stronger markets while allowing for cost recovery and sustainability. The customers are either richer (since subsidies are more modest) or the technologies are sized small enough for poorer customers with limited consumption.

Decentralized Model: These are fully decentralized models in which local actors implement distributed generation technologies. The impact on access and improved electricity service varies greatly. In many cases, the impact of these models is limited due to lack of resources. However, with only an exception or two, sustainability and replicability are medium or high for the models examined. This is due, in part, to the ability of decentralized models to tailor their service to the customer mix and to the need for these models to recoup all or most of their costs due to their limited access to subsidies. There may be limited amounts of donor funding available, but unlike the donor model, the decision-making and implementation in the

---

[1] A more detailed discussion of the results in presented in the appendix to this chapter.

decentralized model is local. The decentralized models can be roughly divided into three categories: first, the well established and independent mini-grids running on fossil fuel (in all three countries) and small hydro-power (in China); second, the established models which serve niche markets such as renewables for productive activities (e.g. Brasil Sustentável – BRASUS in Brazil) or solar home system programs for richer customers (e.g. Khmer Solar); and third, the "new" models that employ either novel technologies or novel business models. These new models typically serve niche markets but have not been around long enough to have a significant impact and allow evaluation of sustainability or replicability.

Figures 6.1, 6.2 and 6.3 show how these different models score on electricity access and sustainability as a function of centralization and operating subsidies. Figure 6.1 compares the models on the criterion of access. Models using well-known technologies that can serve varied customer groups (e.g. the small hydro facilities in China or the diesel micro-grids in Cambodia) provide high rates of access. Decentralized models serving niche customers (e.g. the productive activities supplied by the BRASUS program in Brazil) have lower impacts on overall access to electricity, though their impact on the niche they serve can be quite high. Models using newer technologies (e.g. biomass gasification systems as in the Cambodian cooperative case) or financing schemes (e.g. leasing solar home systems, such as the IDEAAS case in Brazil) have been most limited so far in broadening access, primarily because they have not had an opportunity to be proven and replicated. However, their future potential could be quite high. The mixed models, based largely on supporting key technologies as is done in China, are mid-range in centralization and have been quite successful in improving access. Finally, there are the highly centralized models associated with government ministry, donor or utility efforts. When supported strongly by the central actor responsible, they can result in high levels of access. However, unlike the decentralized models which

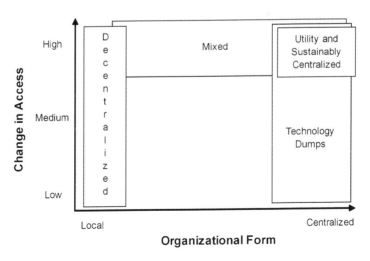

Fig. 6.1 Centralization and access for distributed rural electrification

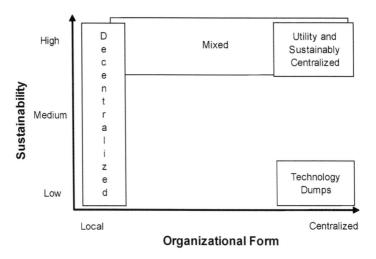

Fig. 6.2 Centralization and sustainability for distributed rural electrification

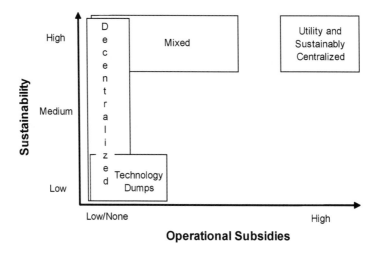

Fig. 6.3 Operating subsidies and sustainability for distributed rural electrification

can and do serve a wide variety of customers, the centralized models tend to be focused on supplying only a basic level of household electrification. Technology dumps, in which technologies are installed in rural areas by a central actor without any ongoing support mechanism, can vary in their impacts on access. Some, such as the dissemination of PV systems in Brazil, have greatly increased short-term access. Others, such as the Japanese PV/Hydro project in Cambodia, have been more limited in their access impacts.

In terms of the sustainability of the model over the longer term, there is also a relationship with the degree of centralization of the model (Fig. 6.2). While there is some variation among the decentralized models, many of them are quite sustainable.

Mixed models, because they have been based on modest subsidies and cost-recovery, also score high on the sustainability metric. Meanwhile, the highly centralized models can be split into those that have some mechanisms for sustainability (such as strong cross-subsidy systems that function well, as in Brazil) and those in which inadequate provisions have been put in place for ongoing operations (i.e. the technology dumps).

The centralized models that can be considered sustainable are dependent upon continued commitments to high subsidies (either direct or indirect), as illustrated by Fig. 6.3. In the case of the decentralized models, on the other hand, the absence of subsidies or a strong central actor from the start compels an emphasis on cost-recovery, even if this means high prices to consumers. The mixed models also generally have low (or zero) operating subsidies and a correspondingly high score on sustainability. There are some exceptions, for example some Chinese small hydropower projects, in which outside pressures have resulted in an inability to recover all costs but the model continues to operate due to support from local or centralized government institutions. Technology dumps, lacking organic local project development at the outset (including incentives for cost recovery) and provision for ongoing support, tend to fail quite rapidly as routine technical failures take equipment permanently out of operation.

An important conclusion of this study is that, in the absence of strong central support for rural electrification, alternative electrification models (e.g. private diesel operators, cooperatives, NGOs providing alternative energy) emerge to meet the needs of different consumers. Lacking financial support from the central government, successful models have had to meet requirements for financial sustainability in other ways. These independent efforts tend to serve a customer base exhibiting the following characteristics:

- Users include productive activities or other high energy consumers (e.g. coops and NGO projects in Brazil)
- Relatively wealthier (e.g. PV customers in Cambodia, wind and hybrid customers in China)[2]
- Willing to pay very high prices for very low consumption (e.g. unlicensed diesel genset customers in Cambodia)

Another point that emerges from the study is that all distributed technologies can be used successfully. There have been and continue to be situations in which technologies are inadequately piloted before wider distribution, are manufactured poorly or have other technological shortcomings. In the great majority of recent cases, however, technology implementations have failed more for institutional reasons than technical ones. In the absence of outside support to introduce renewable

---

[2] These are customers that are at the top of the "base of the pyramid." The base of the pyramid, a term covering the vast majority of the population that is usually ignored by commercial enterprises due to assumptions of their low buying power, has become a powerful organizing idea for creating new opportunities to make money while solving societal problems and meeting environmental goals. See, for example, Hart (2005).

technologies, local technology choice will tend towards diesel generation (e.g. diesel mini-grids in Cambodia and Brazil, battery chargers in Cambodia). Renewable energy technologies have generally relied on regional, national or international institutional support for introduction, product improvement and market improvements (e.g. wind, PV and small hydro in China, PV in Brazil). However, this can become a problem when renewables introduction goes hand in hand with the technology dump approach.

From a policy perspective then, it is clear that central governments and donors have a role to play if increased electricity access is considered a priority. However, that role should be more indirect than it has sometimes been in the past. Modest subsidies designed to spur technology development, favorable policies such as reduced taxes, regulations that protect and don't discriminate against distributed generators are all ways in which governments can help foster and promote rural electrification efforts for the most remote populations. In particular, if distributed electrification models that currently contravene regulatory statutes (particularly mini-grid models) are to be formalized, there must be provision made for the type of service they can provide. Holding them to the same standards as large grids is not feasible. Cambodia's regulation allowing for access that is less than 24 h per day is one good example. That regulatory and policy environment must also be stable. In China, it was only because of the strong support of the central government and the soft budget constraint of the small-hydro stations that they were able to survive the partial takeover of their systems by the state-owned grid companies in the late nineties. Similarly, international donor programs must be geared not towards providing particular technologies but rather in creating access to resources and supporting efforts to create markets (e.g. the World Banks Rural Electrification Fund in Cambodia and international donor aid to Brasus in Brazil).

The implication, however, is that rural residents may be served with electricity that is expensive and/or of low quality and only for certain hours of the day. Some may not be able to pay those prices and be excluded from service. Others may not be willing to pay and equity concerns of having the poorest of the population paying the highest prices may make it politically unacceptable. Thus far, the only solution to that problem that seems to have worked is the Brazilian centralized utility model which allows cross-subsidization to reduce tariffs for low income consumers. This requires strong central government support through regulation and an efficient subsidy mechanism that still ensures utility viability without raising other tariffs high enough to cause discontent or exit from the system. It also implies a lack of diversity and the eventual absence of alternative models that might better take into account local needs (including the need for electricity for productive uses).

One alternative is partnerships between small actors such as NGOs, cooperatives and small entrepreneurs and the utilities within a regulated concession model. This is beginning to occur in Brazil. A second alternative would be a mechanism to allow smaller actors to access cross-subsidization funds that come from consumers of urban utilities. Such transfer payment systems have not been attempted to my knowledge and a host of questions would have to be answered regarding how it would be arranged and protected from capture. Such a scheme would also imply that all players, down to the smallest ones, would be regulated entities raising problems of the transaction costs for both parties. Regulators in the electricity sector are not used to having to deal with a large number of small regulated entities. Similarly, license applications and other regulatory transactions would have be kept minimized and simple in order to these small actors to be able to participate in the system. These questions and issues are addressed in more depth in the following two chapters.

Appendix: Detailed Summary of Results

One key question has been whether distributed electrification should be accomplished through centralized operations or through decentralized actors. The results, as shown in the following tables, indicate that successes and failures can be found in both centralized and decentralized models. Highly centralized models are more likely to result in large increases in access, while more decentralized models are sometimes more limited in their effects as shown in Table 6.4. This is, in part, due to the fact that subsidies, particularly for large capital expenditures, are preferentially given to centralized models, as can clearly be seen in Table 6.5. The centralized models are those have tended to focus on broad electrification programs, geared primarily to households (e.g. the Bahia PV distribution program or the Chinese Township Electrification Program). They also tend to be geared towards more universal service at the lower end of the income scale (including providing electricity for communal structures in low income areas, as in Brazil's PRODEEM). Less centralized options include those targeting richer households that can afford more expensive systems (e.g. those in the Inner Mongolia Autonomous Region), can provide differentiated service for different users (e.g. the Cambodian rural electricity entrepreneurs) or target productive activities (e.g. BRASUS in Brazil).

The relationships between sustainability and centralization and between replicability and centralization show a slightly different picture (Table 6.6 and Table 6.7). Decentralized models are more sustainable and replicable. No decentralized models examined in this study were given a low rating on either count. By contrast, the highly centralized models that are sustainable and replicable are those that are highly dependent upon subsidies.

While some centralized efforts have been successful as a result of this strong support and high subsidies, that is not universally the case. Projects such as the Japanese PV/Hydro system in Cambodia only benefit initially from the high subsidies. It would also be misleading to think that centralized models with high subsidies are the best solution. As will be discussed further in the following chapter, decentralized operations can often meet niche needs and segment the market in ways that improves service over the long term and meets more expansive development needs rather than just rural household needs.

The relationship between sustainability and operating subsidies is shown explicitly in Table 6.8. Similarly, Table 6.9 shows how replicability is related to capital cost subsidies. What these results show is that models designed to recover a significant

Table 6.4 Impact of centralization on access

Access		Low	Medium	High
	High	CA: REE CH: Diesel CH: SHP (private) CH: Small coal	CH: IMAR Wind CH: PV Cash CH: SHP (recent)	BR: Utility SHS CH: microhydro BR: Bahia PV BR: Utility diesel CH: SHP (early) BR: PRODEEM
	Medium	CA: NGO PV BR: Ind. diesel BR: Coops CA: Khmer solar BR: Brasus	CH: IMAR PV/wind	CH: TEP
	Low	BR: IDEAAS CA: SME Biomass	BR: SBC	CA: Japan PV/Hydro
		Low **Centralization**	**Medium**	**High**

Note: This and subsequent tables use the abbreviations BR, CA and CH to indicate models in Brazil, Cambodia and China respectively. In addition, there are a few models for which the available information is more limited (the solar home system diffusion program in Bahia by the World Bank, the Chinese micro-hydro program, small coal in China and independent diesel in Brazil). More detailed entries in the country tables were not made, but enough information is available (primarily through interviews) on their general characteristics to include them in these tables.

Table 6.5 Relationship between centralization and subsidies

Subsidies		Low	Medium	High
	High			CA: Japan PV/hydro CH: TEP BR: PRODEEM BR: Utility SHS CH: microhydro BR: Bahia PV BR: Utility diesel CH: SHP (early)
	Medium		CH: IMAR wind CH: PV cash BR: SBC CH: SHP (recent)	
	Low	CH: SHP (private) BR: Brasus BR: IDEAAS CA: SME biomass		
	None	CA: REE CH: Diesel CH: Small coal CA: NGO PV BR: Ind. diesel BR: Coops CA: Khmer solar		
		Low **Centralization**	**Medium**	**High**

Table 6.6 Impact of centralization on sustainability

Sustainability	Low	Medium	High
High	BR: BRASUS CA: REE CH: Diesel CH: Small coal BR: Independent Diesel BR: Coops CH: PV cash CH: SHP (private)	CH: SHP (recent) CH: IMAR wind CH: IMAR hybrid	BR: Utility Diesel BR: Utility SHS CH: microhydro CH: SHP (early) CH: TEP
Medium	CA: Biomass Coop BR: IDEAAS solar CA: Khmer solar CA: NGO PV		
Low		BR: SBC	BR: PRODEEM CA: Japanese PV hybrid BR: Bahia PV
	Low **Centralization**	**Medium**	**High**

Table 6.7 Impact of centralization on replicability

Replicability	Low	Medium	High
High	BR: Independent diesel CA: REE CH: SHP (private) CH: Small Coal CH: Diesel CH: PV cash BR: BRASUS	CH: IMAR wind CH: IMAR hybrid CH: SHP (recent)	BR: Utility diesel BR: Utility SHS CH: SHP (early)
Medium	CA: Biomass Coop BR: Coops BR: IDEAAS CA: NGO PV CA: Khmer solar		CH: Microhydro
Low		BR: SBC	CA: Japanese PV/ hydro CH: TEP BR: Bahia PV BR: PRODEEM
	Low **Centralization**	**Medium**	**High**

portion of capital costs and cover operations and maintenance are more sustainable and replicable. Modest subsidies that encourage the formation of markets and technology development have enhanced replicability and sustainability (e.g. China's renewables programs). Models that subsidize all, or nearly all, capital and/or O&M costs without also creating a strong institutional support structure have not been

Table 6.8 Relationship between operating subsidies and sustainability

Sustainability	None	Low	Medium	High
High	BR: BRASUS CA: REE *CH: Diesel* *CH: Small coal* *BR: Independent Diesel* *BR: Coops* CH: PV cash *CH: SHP (private)*	CH: IMAR wind CH: IMAR hybrid CH: SHP (early) CH: SHP (recent)		BR: Utility diesel BR: Utility SHS *CH: microhydro* CH: TEP
Medium	BR: IDEAAS SHS CA: NGO PV CA: biomass Coop	CA: Khmer solar		
Low	BR: SBC CA: Japanese PV/hydro *BR: Bahia PV distribution*			BR: PRODEEM
	None	**Low**	**Medium**	**High**
	Operating subsidies			

Table 6.9 Relationship between capital subsidies and replicability

Replicability	None	Low	Medium	High
High	CA: REE *CH: Small coal* *CH: Diesel* *BR: Independent diesel*	*CH: SHP (private)*	CH: IMAR hybrid CH: IMAR wind CH: PV cash CH: SHP (recent) BR: BRASUS	BR: Utility diesel BR: Utility SHS CH: SHP (early)
Medium	CA: NGO PV	*BR: Coops* BR: IDEAAS SHS CA: Khmer solar	CA: Biomass Coop	*CH: microhydro*
Low			BR: SBC	CA: Japanese PV/hydro CH: TEP *BR: Bahia PV Distribution* BR: PRODEEM
	None	**Low**	**Medium**	**High**
	Capital Subsidies			

sustainable and replicable. Examples in which this institutional support mechanism was not provided include technology donations (e.g. the Japanese PV/Hydro project in Cambodia) and the township program in China. Examples in which a significant portion of costs are not recovered but in which strong institutional support exists for continued operation include the centralized utility DG projects in Brazil and many SHP facilities in China.

Distributed models with low subsidies but high sustainability are those that are able to cover their ongoing costs through tariffs or other means. This includes private solutions such as the Cambodian electricity entrepreneurs and the PV cash market in China. Some models have low operating subsidies and low sustainability, indicating that they are not able to cover their costs and do not have the subsidies to maintain their operations. These are primarily failed international donors project such as the Cambodian PV/Hydro Japanese project and the initial attempt to install PV systems in Bahia. On the other end are those that rely almost entirely on subsidies to be sustainable, primarily the centralized utility model of Brazil and centralized efforts in China. These will continue to operate as long as the subsidies are assured. Of course, this does mean that their sustainability can be considered to be more conditional than the models that do not rely on subsidies but are sustainable due to operating cost recovery mechanisms internal to the model. It also raises the question of whether overall results would be better if successful subsidy models could be found that included decentralized actors. This is discussed further in the following two chapters.

Reference

Hart SL (2005) Capitalism at the crossroads: the unlimited business opportunities in solving the world's most difficult problems. Wharton School Publishing, Upper Saddle River, NJ

Chapter 7
Beyond Charity: Universal Service and a Vision for Distributed Electrification

Keywords Policy • Policy reform • Regulation • Tariff subsidies • Universal service

Introduction

The statistic most often used to measure progress in rural electrification is the number of households (or people) with access to electricity. As a result, a lot of effort is expended on changing this statistics by increasing household connections.[1] By focusing primarily on access, electrification can often be limited to the most basic household needs rather than as part of creating the necessary conditions for development. Moreover, it tends to treat all households and all regions as being essentially the same, ignoring differences between customers in terms of both their needs and their resources. These differences include the willingness and ability to pay of both households and non-household customers in rural areas.

The second plank of these universalization programs is generally to keep prices for rural customers low in order to meet urban/rural equity goals. Sometimes this means they should be on par with urban customers, despite the higher costs of rural electrification. Often the prices are actually kept well below urban prices (sometimes to the point of providing the electricity for free). This generally arises out of considerations of equity, that the poorest should not pay the highest prices for electricity.

This focus on household electrification and urban/rural equity requires large subsidies for capital and/or operational expenses.[2] This is accomplished through various means including technology giveaways (e.g. free solar home systems),[3]

---

[1] Often the access statistic seems to be limited to official connections through licensed or legally recognized entities. This ignores independent electrification efforts and systems that are not within the official institutional structure.

[2] This is true of both grid-based and distributed electrification efforts when directed by centralized universal service goals.

[3] This is simply an extension of policies to provide grid connections for free to rural customers when a policy of low prices is in places.

H. Zerriffi, *Rural Electrification: Strategies for Distributed Generation*,
DOI 10.1007/978-90-481-9594-7_7, © Springer Science+Business Media B.V. 2011

cross-subsidies with richer consumers, and soft budgets for government utilities. Similarly, operations and maintenance costs can be covered under cross-subsidies, as in the grid extension case. In some cases, under-priced or even free capital equipment is provided but then operations and maintenance are not supported. In the case of centralized utilities, this can lead to bankruptcy, such as in India, or a limitation on the scale of the program due to lack of resources. System maintenance or expansion become limited by the lack of funds available. They are also limited in their ability to raise tariffs for richer urban and industrial customers out of fear that they will exit the system and install their own generation, further eroding the revenue base (already a major problem in India).

The result is electrification that serves only the most basic needs of the rural poor, a limited ability to recover costs which can eventually stall universalization programs and a push towards centralized solutions both institutionally and technologically when distributed options may be superior. Local solutions outside the utility system have no ability to cross-subsidize their services or be financially backed by the state and so must charge higher prices. However, it is difficult to justify such investments when the centralized competitor is essentially giving away power either through the grid system or through their own distributed generation programs. This reduces incentives for investment in alternatives that might better meet rural electricity needs.

The high subsidies for rural grid extension and centrally planned universal service programs can also feed into normative concepts about the inferiority of distributed generation versus grid power. Normative biases against distributed systems in favor of the grid result in distributed rural electrification being seen as "second-class" electricity. This technology difference is seen as an issue of equity and fairness, and can result in a refusal to install distributed systems based on the belief that this would preclude the future installation of the "superior" grid technology. Of course, a comparison of distributed rural electrification to urban grid systems is not fair. By highly subsidizing rural grid extension, these programs mask the true comparative costs and benefits.

The next section examines how these distributed electrification models meet these twin goals of access and equity. For each country, universal service goals have been stated and institutionalized to differing degrees. Similarly, equity goals differ within all three countries. The result of how these differing goals are met in each of the three countries provides evidence of the need for new mechanisms to foster distributed electrification models that contribute to rural development.

The third section of this chapter then provides an alternative vision of distributed electrification. This vision is also based on the goal of expanding access, but doing so in a way that is sustainable over the long term and contributes to local economic development. The final section outlines some of the key changes that need to be made in order to move forward with such a vision. These include changes to the regulatory system and a rethinking of the social contract between governments and the people they serve.

The Old Vision: Cheap Service for All

Universal Service

There are currently roughly 4.3 million people in Brazil that do not have access to electricity. The utilities have had little incentive to meet more than basic electrification goals. It was not until a combination of regulatory action and financial carrots forced them to act that the privatized utilities began aggressively electrifying rural areas. The Luz Para Todos program is the latest in a series of Brazilian government programs with the goal of improving electricity in rural Brazil. The explicit goal of the program is universal electrification with a fixed date to meet the program's objectives:

> Art. 1o – Fica instituído o Programa Nacional de Universalização do Acesso e Uso da Energia Elétrica – "LUZ PARA TODOS", destinado a propiciar, até o ano de 2008, o atendimento em energia elétrica à parcela da população do meio rural brasileiro q ue ainda não possui acesso a esse serviço público (Institui o Programa Nacional de Universalização do Acesso e Uso da Energia Elétrica 2003).
>
> Art. 1 – A National Program for Universalization and Use of Electrical Energy – "Luz Para Todos" shall be instituted, destined to provide by the year 2008, electricity service to the portion of the population in rural areas of Brasil that still do not possess access to this public service (Author's translation).

The program provides funding to the Brazilian utilities to meet their regulatory obligations to serve everyone within their exclusive service territory. While it is becoming increasingly clear that this target will only be met in the more industrialized south, where the challenges are not as great, the government remains committed to the goal.

The targets have, therefore, been primarily households. This can be seen in COELBA's PV program, which it is using to meet its obligations in the more remote regions of Bahia. The PV systems are for households only and there is no programmatic goal to include other customers or to provide more than basic household electrification. Even in the Amazon, diesel minigrids are undersized for inclusion of productive activities. For example, in Aquidabam, the diesel generator owned and operated by the central utility provides enough electricity for households, community structures, and a few stores. However, even then the system is unable to meet all the demand. Refrigeration for the agricultural product, which would improve the community's ability to get its product to market, cannot be met, leading to an outside project for a biomass gasification system.

Previous centralized programs, such as the PRODEEM, have had a focus on community structures, such as schools or health clinics rather than households. If sustained, electrification of these services can clearly impact human welfare in rural areas. The argument has been made that even improvements in education and health can be considered "productive" uses of electricity (Cabraal et al. 2005). However, in terms of contributing to improved economic opportunities, programs like PRODEEM do not have a direct effect.

The limited progress centralized utilities have made in rural electrification has lead to a number of alternative distributed models. In contrast to the centralized utility model, these alternative models sometimes go beyond basic electrification. Some are focused on providing electricity to productive activities in order to improve economic output and development (e.g. BRASUS). Others remain focused on households (e.g. IDEAAS), but allow for higher levels of electricity consumption than the basic levels provided by the utilities and are decentralized in both technology and organization.

These alternative models have been limited in Brazil. They cannot compete directly with the centralized utilities both because of the legal mandate and because of the tariff structure and subsidy system to keep rural prices low (dealt with in the next section). The recent expansion of the centralized system as the result of the Luz Para Todos program calls into question the role these alternative models can play in future rural electrification efforts. To a certain degree the Brazilian government has recognized that the focus of the centralized utilities on basic household electrification is limited and started to develop integrated action plans to meet more general economic development needs. These action plans would utilize more distributed actors rather than the centralized utilities.

It should be noted that the mandated service territories does not prevent the sales of individual electricity generating units, such as solar home systems. However, as with the other alternative models, the incentives to purchase such systems are affected by the expansion of the utility system at little cost to the consumer.

Cambodia is perhaps the direct opposite of Brazil. This is not to say that the Cambodian government would not like to see all of its rural population electrified. That is clearly a goal over the long term with a target of 70% of the population getting grid-quality electricity by 2030.(World Bank 2003) However, most documents simply discuss the low level of grid electrification in Cambodia and plans for rural electrification without reference to an absolute timetable.

This is because the situation in Cambodia is significantly different than in Brazil. First, Cambodia does not have the financial resources to contribute significantly to the costs of rural electrification. While the government can set into place mechanisms to improve rural electrification (e.g. regulations, international donor projects, inter-connections with neighboring countries), it cannot establish the type of ambitious programs that Brazil has because it does not have the resources to invest. Second, in the absence of centralized action in the past, individual actors have moved in to fill the gap by setting up mini-grids and battery charging stations. Thus, while nearly 25% of the Cambodian population officially has electricity, the reality is that nearly 100% of the Cambodian population has a light bulb.

Rural electrification in Cambodia is thus not the result of official government policy and support, but the result of actions at the local level. The result has been universal service in which most households receive only the bare minimum of electricity, as in Brazil. This low level of service is the result of the economic situation of the rural population and their inability to afford more electricity, especially given the high prices the Cambodian entrepreneurs must charge. However, the difference with Brazil is that those that are able to afford more, such as richer households or productive enterprises, are able to obtain the higher levels of service they desire. This is either through the rural entrepreneurs or by exiting the system and purchasing their own electricity generation equipment (e.g. solar home systems or their own diesel generator). Local action has meant actors have been able to adjust to local needs.

In many ways China sits between Brazil and Cambodia in terms of meeting universal electrification goals. In the early years of distributed rural electrification, the goal was meeting agricultural and industrial production needs. While much of the program was centrally mandated and funded, there was significant local involvement and benefit from rural electrification in the fifties, sixties and into the seventies. However, providing household electrification was secondary to meeting the production needs of the state.

Within China's more recent 5 year plans, and the programs of the National Development and Reform Commission, are the goals of providing electricity for rural populations. In part this is to reduce the disparities that exist between the more industrialized eastern and more rural western parts of the country. To that end, the Chinese government has gone beyond setting a simple access objective and established criteria for considering rural counties to be "electrified." This includes a minimum consumption level of 200 kWh per person (Tong 2004).

These goals have been met in a variety of ways. Central planning is continued in programs such as the Township Electrification Program and the microhydro to replace fuelwood program. As with other centrally managed programs the focus has been on meeting very basic household needs. This is true even when the choice has been to install village level grids, as in the township program.

The primary model for distributed rural electrification in China has been the centrally supported, but locally owned and operated, small hydro utilities. The small hydro power in many counties has been combined with small coal plants to take advantage of cheap local resources and to meet demand during times of low water availability. These local utilities provide electrification for entire communities, including households, community structures and productive activities.

The other distributed rural electrification models are geared towards meeting household needs. The wind and wind/pv hybrid systems sold in Inner Mongolia allow richer households to meet higher electricity needs. The modular PV systems sold in western China allow people to expand as incomes allow. Both of these models have only involved the central governments in supportive roles that encourage technology and market development.

The universal service policies and outcomes for all three countries are summarized in Table 7.1.

Urban/Rural Price Equity

Brazil's distributed rural electrification system has three main characteristics as a result of the regulatory regime and policies of the central government:

Table 7.1 Universal service policies and outcomes for the three case study countries

Country	Policy	Implementation	Outcome
Brazil	Universal Service with target date and funding	Centralized utilities	Basic household electrification, limited outside actors
Cambodia	Eventual universal service	Distributed utilities	Segmented markets, near-universal service if include battery chargers
China	Universal service with criteria for "electrified counties"	Local government utilities	Overall community electrification
		Technology development	Promotion of markets for niche markets
		Centralized programs	Basic household electrification

1. Exclusive service territories for the utilities with a requirement for utilities to provide electricity to all consumers within their territory (dealt with in the section above).
2. Subsidies by the central government to utilities for capital costs under the Luz Para Todos program.
3. Mandated low tariffs for low income and rural customers.

The second and third planks of governmental rural electrification efforts in Brazil establish a goal of price equity between rural and urban areas. In theory, under this system, poor rural (and urban) consumers pay tariffs that are commensurate with their much lower incomes.[4] The tariffs charged to rural customers are significantly below what is required for cost recovery, even accounting for the subsidies for capital costs. In order to be able to charge such low tariffs, the privatized utilities charge their urban customers higher tariffs and cross-subsidize the rural consumers (e.g. COELBA in the state of Bahia). The utilities still owned by government holding companies (such as those owned by Eletronorte) have soft budget constraints that allow them to show losses. These losses are covered by Eletronorte's other business units. In effect, this implies a cross-subsidy by the customers of those other business units, primarily the privatized utilities. In the one case cross-subsidies are internal to the business. In the other, the cross-subsidy is shifted out of the distributed electrification business.

Along with the exclusive service territory regulations, the mandated low tariffs for rural customers create problems for alternative distributed models. Even if the regulations were changed to allow independent mini-grids to operate in rural areas, the customer base would not allow cross-subsidization and it would be impossible to charge the low tariffs mandated by the regulation. For those distributed models

---

[4] The situation is complicated by the fact that qualification for the low tariff is based not on income but on consumption. Low consuming households are assumed to be also low-income households. There are no corrections made for various factors that could skew the correlation between consumption and income (such as household size).

based upon sales or rentals of individual units (such as solar home systems), there is no prohibition against those businesses. However, the incentives for customers to purchase the systems are reduced when the option to receive service from the utility at highly reduced rates is available (or is expected to be available soon).

As mentioned in the previous section, Cambodia's economic situation does not allow it to have a policy of reducing rural tariffs. Achieving parity between urban and rural tariffs is not possible through either government or utility subsidization. Reducing prices in rural areas below urban areas is even further from the possible. There are, however, some centralized models in Cambodia that do highly subsidize a limited numbers of installations. The Japanese PV/hydro project is an example in which capital costs have been completely provided for by a donor through the central government ministry.

The rural electricity entrepreneurs have to charge tariffs sufficient to recover their costs. Subsidies are not available for capital costs or operating costs. For the operating costs, some implicit cross-subsidization does occur even on the most basic level. The internal accounting of some REEs is not very sophisticated, such that costs for running the system is not always properly allocated between the different classes of customers (mini-grid households, battery charging households and productive enterprises). However, unlike the Brazil case, the cross-subsidization is between rural customers and is not explicit. Despite any minor cross-subsidization, all customers pay extremely high rates due to the high costs with rates commonly in the range of 50 cents/kWh.

The situation in China is much more complicated than in either Brazil or Cambodia when it comes to the issue of urban/rural equity in electricity prices. It is only in those distributed systems that have been hooked up to the expanding grid system that a formal policy of price equalization exists.[5] The policy is not for significantly lower prices in rural areas, as in Brazil, but for price uniformity across a grid system. Aside from these cases, the policies for the small hydropower utilities is a classic cost-plus accounting with tariffs set to recover those costs. However, while that is the policy in theory, in practice there are other factors that come into play:

1. The Chinese government's goals for rural electrification to contribute to rural development and reducing the rural/urban divide has led it to implement a number of policies that reduce the cost of rural electrification (e.g. small hydropower systems have favorable tax treatment and receive low interest loans from government banks).
2. Many of the financial liabilities are shifted onto the central banks due to low loan repayments.
3. Inefficiencies in the small utilities, local political pressures, and a host of other factors result in some utility tariffs remaining below the cost of supply.

---

[5]This has been a common occurrence as the grid system reaches areas previously served by local small hydropower utilities. The massive expansion of the grid in the late nineties resulted in many distribution systems being absorbed into the large grid network.

The result is that while there is no policy to force rural utilities to charge rates equal to or below costs, the larger policy goals of the Chinese government, the dysfunctionality of the banking sector and a host of other factors result in lower prices for many small hydropower utilities.

A few of China's distributed electrification models explicitly include high subsidies for capital and/or operating costs. The township electrification program's capital costs were completely underwritten by the central government. Operating cost recovery in the first 3 years of the program has varied widely between program sites with some consumers paying no tariffs during the initial phase and others paying relatively high prices. It is unclear what will happen to tariffs as the program moves into the post-implementation phase. Another program, to replace the use of fuelwood with micro-hydro, is also highly subsidized, particularly in its capital costs. Both of these programs are highly centralized. While they are not explicitly designed to achieve close price equity between urban and rural areas, their goal of improving the economic situation for rural consumers and the way they are implemented does result in low prices for some consumers.

The renewable energy market programs stand in contrast to both the highly centralized and subsidized models and the implicit subsidies of the small hydro utilities. One key to the success of both the PV cash market and the IMAR wind market was deliberate programs of technology development and diffusion. Rather than supplying technology centrally at low cost as in the later TEP program, these programs used target subsidies to support market mechanisms. Ultimately, customers are financially responsible for the purchase as well as ongoing maintenance.

The policies and outcomes of the three countries when it comes to price equity are shown in Table 7.2.

Table 7.2 Price equity policies and outcomes for the three case study countries

Country	Policy	Primary implementation	Outcome
Brazil	Below-cost and below-urban prices for rural customers	Centralized utilities	High subsidy burden, exclusion of other actors
Cambodia	Cost-recovery	Distributed utilities	High tariffs but costs recovered, highly sustainable and replicable, limits to consumption at lower incomes
China	Mixed	Local government utilities	Cost-recovery tariffs in theory, shift of burden to banking sector, push for equalized urban/rural tariffs after grid connection
		Technology development	Modest capital subsidies to spur technology
		Centralized programs	High capital subsidies, varying operating subsidies

Impacts of Universal Service and Equity Programs

In Brazil, a policy of basic household electrification has been institutionalized through the Luz Para Todos program with only recent and limited moves towards more integrated actions. Alternatives that are focused on productive activities or on households that need higher levels of power are marginalized by the system. By contrast, the organic growth of the rural electricity systems in Cambodia has resulted in the varied needs of rural customers being met, albeit at high costs and poor quality due to a lack of resources. Meanwhile, China has a varied experience in which central government supported programs have tended to reflect local needs and the ability and willingness of customers to pay for power. However, those programs run by the central government (rather than just supported from the center) have tended to follow the path of basic household electrification.

In Brazil the effect of these two policy priorities (universal access and cheap power) can be seen in regulations that keep prices low (requiring cross-subsidization) and mandate exclusive service territories for utilities and in government subsidies for capital costs that can only be accessed by those utilities. In China, where rural electrification did promote local solutions and electricity for production, a recent shift towards universal access led to a program focused solely on household electrification. While it did use distributed generation, the Township Electrification Program's top-down implementation did not create the institutional arrangements necessary for sustainability. As another example, India's utility system suffers from insufficient tariffs and high losses and has been bankrupt for a long time. In the end, rural electrification efforts fail to live up to the needs because utilities view it as a liability while proponents (including within the international community) view it too often as charity.

The impacts of policies for universal service and price equity are shown in Fig. 7.1. The priorities lead to a focus on basic household electrification and on reducing prices for rural consumers to well below the cost of service. This leads to a dominance of centralized mechanisms for meeting these policy goals. The results can include bankruptcy of utilities, reduction of incentives for alternative models that might meet different goals and needs and stagnation at the level of basic electrification.

An Alternative Vision for Local Electrification

If the traditional policy focus on universal access and cheap power leads to an over-reliance on centralized solutions that don't meet the needs of rural populations and are often unsustainable, what is the alternative? A new vision for rural electrification must account for three important factors:

- Electrification must be linked to larger development goals. This is not to say that household electrification must be abandoned. However, electrification efforts that move beyond basic household electrification have a greater chance of being

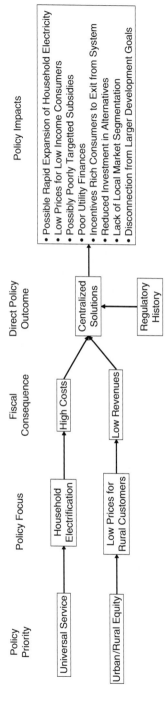

Fig. 7.1 Impact of policies for universal service and price equity between urban and rural consumers

both successful and of improving welfare in rural areas. The importance of linking rural electrification specifically to productive activities is discussed below.[6]

- Distributed electrification efforts must be de-linked from the equally important and overlapping priority of solving the problem of global climate change. While there are clear links between solving rural energy poverty problems and the global climate change problem, there are also serious downsides to linking the two so closely that only renewable energy technologies are considered for meeting rural energy needs. This issue is discussed further below.
- Distributed electrification efforts must be adaptable in order to meet local conditions and account for differences in needs and resources among rural populations. If allowed, distributed electrification models can segment the population according to the demands, willingness to pay and ability to pay of various rural customers. This issue is discussed further below.
- Distributed electrification efforts should take advantage of the de-coupling that is possible between technologies and institutions. Small scale technologies mean that organizations ranging from the smallest (households) to the largest (national governments) can be involved and institutions must be reoriented to accommodate that fact.

If the focus is on truly meeting the needs of end-users, this will require more than simply money and the provision of technologies (Barnett 1990; Barnes and Halpern 2000). In order to avoid the trap described in the previous section, it will be necessary to move away from a traditional donor and government-centric model. The evidence from cases in all three countries included in this study is that there are numerous business models that can be effective in providing electricity in rural areas. In the absence of sustained financial support from the central government or richer consumers (e.g. in Brazil), successful models have had to meet requirements for sound financial sustainability in other ways. These independent efforts serve customers that exhibit the following characteristics:

- Productive activities or other customers with higher consumption are included in the customer mix (e.g. coops and NGO projects in Brazil)
- Relatively wealthier (e.g. PV customers)
- Willingness to pay very high prices for very low consumption (e.g. unlicensed diesel genset customers)

It is important to note that there is no single technology that has led to the success of the business models. There are various technology options that are appropriate, depending on local conditions and the goal of the distributed electrification effort. There are also a variety of business models used, from cooperatives to dealer models to rural entrepreneurs to local government utilities. In fact, that is the important lesson, there is a dizzying array of possible arrangements and solutions that can be

---

[6] This is not to say that meeting other welfare enhancement goals (e.g. improved education and health) are not important. However, the linkages between electrification and income generating productive activities are often much cleaner and clearer than the linkages to those other welfare measurements.

successful, but they must account for the local conditions, meet the needs of the consumers and match the institutional particularities of the particular country.

In many ways the vision articulated here is a micro-version of some of the ideas in William Easterly's recent book on foreign aid. (Easterly 2006) In his book, Easterly decries the use of "Big Plans" to solve the ills of the developing world and instead articulates a vision of empowering individual searchers (including within the aid community) to find solutions to specific problems. Similarly, this book argues that the big plans for rural electrification have not lived up to their potential (as evidenced by the number of people who still do not have electricity access and the unsustainability of many rural electrification efforts that have been attempted). The solution is to open the field up to a variety of searchers who will find solutions best suited to local conditions. As argued in this chapter and the next one, this does not eliminate the role of governments and donors in solving the rural electrification problem, but it does mean that largely centralized solutions are not the only solution and should compete with other decentralized solutions that may serve rural needs better (Barnes and Floor 1996). This is a highly messy process of experimentation, market building, and institution strengthening. For the same reason that broader markets and economic growth cannot be created through large top–down planning, rural electrification that truly solves the rural energy problem cannot be achieved through highly centralized top down planning and implementation. Experimentation that rewards innovative and simple solutions that match local conditions will move rural electrification beyond a simplistic emphasis on basic connections. This also allows for the important feedback that is necessary to distinguish between good and bad rural electrification projects and to channel resources away from failures and towards those that have more chance of success (Barnett 1990; Easterly 2006).

This emphasis on local needs and local conditions is also what distinguishes this vision from recent sustainable development literature that is focused on the "Bottom of the Pyramid." The Bottom of the Pyramid is a relatively recent term covering the vast majority of the population that is usually ignored by commercial enterprises due to assumptions of their low buying power. This has become a powerful organizing idea for creating new opportunities to make money while solving societal problems and meeting environmental goals. The fundamental argument underpinning the Bottom of the Pyramid literature (e.g. Hart and Prahalad's work) is correct (Hart 2005; Prahalad 2005). There are indeed markets that exist in meeting the needs of those at the bottom end of the income scale. Innovative ideas that match products to the needs of the poor can be successful, including in the field of energy. Where this book diverges from that vision is in its application of sustainable development concepts. Starting with Hart's book, the BoP literature has generally promoted the concept that there are no tradeoffs between economic development, social improvement and the environment. The three pillars of sustainable development can be met simultaneously without any losses. While every win–win situation should be exploited in order to avoid making unnecessary tradeoffs, this does not mean that tradeoffs have vanished simply because those at the bottom of the pyramid are recognized as consumers. This is discussed further below.

In summary, this is a needs-based vision that starts with solving the rural electricity problem by trying to empower all those that can provide solutions and move away from ineffective centralized planning solutions. Again, in the words of Easterly, it envisions providing the necessary support to searchers rather than planners.

Linking Rural Electrification to Productive Activities

The linkages between rural development and rural electrification are complex and varied, but one way in which they are clearly linked is through the support for rural productive activity, often agricultural activity (Barnes 1988; Goldemberg and Johansson 1995; Allderdice and Rogers 2000; FAO 2000; van Campen et al. 2000; Fishbein 2003; Meadows et al. 2003; Ramani and Heijndermans 2003; Cabraal et al. 2005; Cherni et al. 2007). Rural electrification is seen as one of a number of enabling conditions necessary for the development of rural economies and its ultimate success depends on the package of conditions being present that foster growth. This includes both other physical infrastructures (e.g. water supply and transportation links) and social infrastructures (e.g. education and health systems) (Barnes and Floor 1996).

While Barnes and Floor focus on agricultural production, the "conditions that favor income growth" need not be limited to agriculture. Small cottage industries and other rural commercial enterprises can also promote income growth (Martinot et al. 2002). Furthermore, with distributed power generation technologies, the electricity business itself can potentially contribute to growth. However, as with agricultural production, this will depend greatly upon the supporting conditions for economic development.

All three countries in the study show the impact that electrifying productive activities can have on outcomes. In Brazil, the BRASUS program is geared towards providing agricultural producers with electrification and the financial structure shows the emphasis placed on cost recovery and sustainability. Cambodia has numerous rural enterprises across a range of sizes and sectors, including agricultural processors (e.g. rice mills), ice factories, small shops, and welders. These can all benefit from improved electricity supply. The Chinese small hydro utilities are perhaps the clearest example of linkages between rural electrification, productivity and economic growth. Originally, electrification was very specifically geared towards meeting national goals for agricultural and industrial production. Later, as the focus shifted to include households, the importance of electricity for rural production was not lost. Indeed, it is hard to see how the township and village enterprises that form a large part of the rural economy and even of China's national economy could have developed without the supporting electricity infrastructure.

It should be noted, however, that the use of electrification to support rural agricultural activity can also lead to problems if the same policy priorities for household electrification are extended to productive activities. This is best exemplified by the situation in India with regard to agricultural pumpsets. As part of a policy to support rural farmers, pumpsets have been installed without electricity meters and farmers receive free electricity. The cross-subsidies obtained from charging urban industrial and commercial customers higher rates are insufficient to cover the losses and this, combined with other losses, has led to the crisis that has gripped much of the Indian electricity sector for years (Tongia 2007; Dossani 2004; Thakur et al. 2005; Government of India 2006; Int.Don5 January 2006).

De-linking Rural Electrification from Climate Change

The problem with the drive for energy solutions that **must** be environmentally friendly at the same time as serving the needs of the rural poor (both from a social and economic perspective) is that very real tradeoffs **do** exist. These must be addressed and means of evaluating them must be developed. Otherwise, money will keep being allocated to projects that meet an a priori expectation regarding environmental performance. The tradeoffs are then either unacknowledged or essentially dealt with as a residual problem. The goal is reduced to maximizing economic development or social welfare outcomes **given** that the technology chosen protects the environment. Evaluating the tradeoffs ahead of time would lead to a better match of needs and technologies and likely improved overall sustainability. There is also a difference between meeting goals related to the local environment (e.g. air or water pollution) and those related to the global environment (e.g. climate change). The concern here is specifically with tradeoffs related to global environmental change rather than local environmental impacts. To be clear, implementing rural renewable energy projects in these areas could potentially be justified on climate change grounds for any of the following reasons:

Rural Contributions to Climate Change: The disparities in emissions between the OECD and the rest of the world's economies (in absolute, per capita, and historic terms) are a fundamental issue. As the largest emitters, the United States and China are responsible for about 35% of emissions (World Resources Institute 2009). However, Chinese per capita emissions are more than six times lower than U.S. ones and cumulative per capita emissions in China are roughly eight times less than the United States. Overall, world per capita emissions average 5.5 t of CO_{2eq}/ year while the Kyoto Protocol Annex I country average is 14 t CO_{2eq}/year and the non-Annex 1 average is 3.2 CO_{2eq}/year, with the least developed countries emitting just 1.6 CO_{2eq}/year per capita (World Resources Institute 2009). Certainly, at a national level, the vast majority of developing economies have not been a major contributor to the problem. However, this data masks the disparity between urban and rural populations within the developing economies as the data on greenhouse gas emissions is aggregated at the country level (Chakravarty et al. 2009). For example, energy consumption in the industrial centers of Guangdong and Shandong in eastern China are 3–4.5 times what they are in Gansu, a region in the west of the country (data from the Chinese Statistical Yearbook of 2007).

Climate Impacts of Solving Rural Energy Poverty: It is somewhat easier to determine the potential climate impacts of solving the rural energy poverty problem. Certainly solving the basic energy access problem would seem to have marginal impacts on global fossil fuel consumption or on climate change in comparison with the benefits. One estimate shows that eliminating the electricity access problem (i.e. providing electricity to all 1.5 billion people without access at the moment) would result in an increase of CO_2 emission of just 0.15 GtC/year (~2%). Similar results hold for providing a basic amount of propane to all 2.6 billion without access to modern cooking fuels another major energy poverty problem.[7] The IEA estimated that providing another 1.3 billion people with access to LPG by 2015 would result in less than 1% change in global oil consumption (International Energy Agency 2006). This indicates that solving the most basic rural energy poverty problems do not significantly increase greenhouse gas emissions. In fact, there is some evidence that the incomplete combustion of biomass in cookstoves may be a significant source of black carbon, carbon monoxide and other products of incomplete combustion that, in fact, have significant climate change implications (Bond and Sun 2005). If that is the case, then switching from biomass to fossil fuels may in fact reduce net GHG impacts rather than increasing them.

Potential for Learning by Doing: The potential benefits of learning by doing would include the potential benefits of increased markets resulting in greater manufacturing of renewable energy technologies and, thereby, a reduction of production costs as manufacturers improve their production processes.[8] Learning by doing could also result in cost reductions on the implementation side as lessons are learned and institutions are improved to support renewable energy use. The former justification is questionable given the significantly higher installations and demand in the industrialized economies. As of 2006, total renewable power capacity (excluding large hydropower) was 207 GW. Of that, 88 GW was in the developing countries with China (52 GW) and India (10 GW) accounting for the majority of that capacity. More importantly, manufacturing is centered in the OECD economies plus India and China (REN21 2008). The second justification is reasonable in the context of rural energy and a valid argument for support of institutions and market development. However, given the problems that have occurred in implementation of numerous renewable energy projects in the past, it is not clear whether learning is occurring – or, if it is, whether that learning is transferred between projects (Nieuwenhout et al. 2001).

While both of these reasons may be valid under some circumstances, it is also clear that they are not valid under all circumstances. It may be desirable to find those win–win situations, but it is equally important to avoid the lose–lose situations and that is not always the case (Zerriffi and Wilson 2010).

---

[7] Calculations by Robert Socolow of Princeton University based upon 50 W/capita at the current global mix of electricity generation and 35 kg propane per capita.

[8] In this case, cost reductions could also come simply from increased economies of scale due to increased production.

Segmentation of Local Markets

The presence of Distributed Rural Electrification Models that are unsubsidized or only moderately subsidized are important for demonstrating how the needs of rural consumers are met in the absence of central mandates on either the types of service or prices. The Cambodian entrepreneurs, the Chinese renewables markets, and the IDEAAS model from Brazil are all examples of models in which segments of the population get service based on their specific demand and ability and willingness to pay. The Cambodian electricity entrepreneurs serve a variety of customers, including shops and other income generating enterprises. As noted before, these all-day customers help improve the economics of the mini-grids, allowing them to run their generators for longer hours (see Box 1 on market segmentation in Cambodia).

Prices are generally much higher than under the centralized utility system, but demand is generally met. These high prices paid by poor rural populations is usually bemoaned and are used as a justification for centralized programs and donor programs even in places where the rural population is already receiving service. However, these high prices and differentiation in service actually demonstrates that even at the lowest income levels, market segmentation can occur and service can be tailored to fit local conditions. This improves chances of finding viable business models since there is not a one-size-fits-all model that is suitable in all contexts.

Programs to ease the burden of high prices, unfortunately, often do not try to preserve this market segmentation and differentiated service. They can end up destroying viable markets in favor of uniform service and pricing. The challenge is finding ways to meet equity goals without destroying the incentives for actors to actually provide the service best suited to meeting local needs. Cross-subsidies, as traditionally implemented, are problematic because they require a mix of customers that don't exist on the local level. Large and direct subsidies also create problems in terms of continued budgetary support, destruction of markets, reduced incentives, etc.

Local Scale Solutions

Ultimately, the "viability of many small-scale energy technologies (such as stoves and biogas plants) depends on factors that are specific to a particular location." (Barnett 1990) In traditional centralized generation systems the scale of the technology and the scale of the organizational actors are coupled in many ways. The large scale of generation, transmission and distribution systems requires both the operation and regulatory oversight of these systems to match their scope. The one exception, perhaps, is the ability of small local utilities to purchase power from the grid for distribution to end-use customers. Even then, however, the coupling of the small utility to the larger grid requires some form of overarching regulatory structure and at least a part of the industry to be organized at a large scale.

With distributed power generation, it is possible to decouple technological and organizational scale and allow for both large and small scale organizations. It neither requires large-scale organizations to oversee generation and transmission to multiple end-use distributors nor does it require small scale organizations to implement and regulate local production of electricity. Various size-scale combinations are possible for both regulation and implementation of distributed power systems. This allows distributed rural electrification efforts to provide the advantages usually associated with decentralization. The oft-stated general advantages of decentralization (usually considered in terms of either political decentralization and/or fiscal decentralization) are the ability to account for local preferences, improved public participation and good governance (Litvack et al. 1998; Shin 2001).[9] Decentralization is not just a theoretical possibility. Political and fiscal decentralization, as well as the decentralization of infrastructure services is a phenomenon that has been on-going for a while (Bird 1994; Estache and Sinha 1995; Litvack et al. 1998). An example of this is the recent amendment to the Indian constitution that make electricity supply a responsibility of the local Panchayats (rural governmental bodies) in addition to the state and federal governments (Government of India 1992).

In the cases studied here, decentralization has often not been the result of a concerted and planned effort to devolve authority to the local level. Rather, it has been the result of local actors moving in to provide a service where the central government or its agents (e.g. the large utilities) have been absent. This includes local entrepreneurs (most notably in Cambodia), who have been able to judge the needs and capabilities of potential clients to pay for their service and met that need. Non-governmental organizations, both domestic and international, have also engaged in projects worldwide.

One way to account for local factors is, of course, to have the local community itself involved in ownership and operation of the system. Whether it is through the local government, NGOs or cooperatives, organizations made up of community members can be effective in assessing local needs and resources. This has been widely advocated, though not always successfully. Instead, emphasis has often been on centralized grid extension or on the individual household or consumer (Barnes and Floor 1996). One advantage of community ownership and management is that it is possible to create a vested interest within the community for the successful, sustainable operation of the system. However, this is not guaranteed and it may be necessary to require that "the community should make a substantial contribution to the installation and management costs. If this is done, there is a reasonable prospect that the scheme will be responsibly managed" (Foley 1992).

One place where local government involvement has been extensive is China. The relationship between local action and centralized state action on small hydropower

---

[9]The debate about the relative benefits and costs of decentralization and the conditions for successful decentralization, which covers a much wider space than just the issues related to infrastructure provision, is ongoing and beyond the scope of this paper (Shin 2001).

has evolved over the last 5 decades, but at all stages the local community was involved in the development of local hydropower (Pan et al. 2006). In Cambodia, the cooperative structure used to install and run a biomass gasification project ensured community involvement. This was critical to their success in overcoming unforeseen problems and allowed them to raise the tariff when necessary (something that was not possible with the Japanese PV-Hydro project).

Decentralization of both decision-making and service delivery does have some potentially negative implications:

1. There is often a lack of technical capacity at the local level to install systems, and just as importantly, to maintain them. The remoteness of some populations increases the need to have members of the local population trained in at least simple maintenance procedures. "The availability of effective repair and maintenance services, in the final analysis, may well be the principal factor determining whether particular community managed schemes succeed or fail." (Foley 1992) This is, in fact, true of any scheme, not just community managed programs. Technologies that fail in a short period of time due to poor maintenance programs are not only a waste of scarce resources; they can create strong biases against certain technologies that can hinder later efforts.
2. The imbalances in influence and power on a local level can result in the benefits of electrification being captured by local elites. This is discussed further in the next section.
3. Decentralization can cut local activities off from sources of revenue and/or subsidies (including cross-subsidies from richer urban customers), making cost recovery dependent entirely upon lower income consumers. This issue is addressed in the next chapter.

Box 1 Segmentation of Customers by Rural Electricity Entrepreneurs in Cambodia

At first glance, the electricity service provided by the typical rural electricity entrepreneur in Cambodia would simply seem to be poor quality electricity at high prices, an undesirable outcome. In fact, many REEs have been able to provide the customers with the level of service they need and can afford. During the day, the diesel generators can be run to recharge batteries for customers that are not close enough to reach with the mini-grid. At the same time, some power is sent to users that need power for longer periods of time. These are generally customers running small enterprises such as stores. They need to have power and have the extra income to afford power throughout the day. In the evening the REE distributes power over the lines to the houses connected to the minigrid. In this way, the REE can meet the needs of three different customer classes using the same equipment.

Moving Forward

Distributed generation offers the possibility of maximizing the advantages of both technical and institutional decentralization. However, achieving that vision of coupling rural electrification to larger development goals and taking advantage of the opportunities for market segmentation will require major shifts in the way that rural electrification policy is articulated and implemented. These shifts have to occur in both the institutional structures that govern rural electrification efforts and in the financial models used for distribution rural electrification.

The changes that have to be made in the institutional structures that govern and constrain distributed electrification activities can be roughly divided into four:

1. Aligning Macro-level Policies With Desired Outcomes
2. Reconceptualizing Infrastructure Decentralization
3. Creating Flexible and Appropriate Regulations
4. Restructuring the Financing of Rural Electrification (dealt with separately in the next chapter)

Changing Macro-Level Policies

It is well recognized by both academics and policymakers that while individual technology choices are made on the micro-level (that is, by an individual or small group of decision-makers), those decisions are influenced by policies set at the macro (generally government) level (Stewart 1987). The impact of macro-policies on technology choice is the result of changing "firm objectives, resource availability and cost, markets, and technology." (Stewart 1987) The focus, however, is almost exclusively on how policies affect technology choice. The following quote from Barnett is a perfect example.

> The successful pursuit of policies to introduce more efficient energy conversion technologies will inevitably require the rationalization of macro policy not only with regard to policies for prices and subsidies, but also with the other policy instruments which determine the environment in which people choose technology (Barnett 1990).

Little consideration, on the other hand, has been given to the question of how those macro-policies affect the types of organizations rather than the technologies chosen. If prices and subsidies are rationalized in order to introduce more efficient energy conversion technologies, what affect does this have on the types of organizations that can participate in the market? For example, macro-policies to keep rural prices low often result in regulations that rely on cross-subsidies between customers of a single company. This makes it difficult for small entrepreneurs to have a successful business.

We propose therefore a shift in the policy priorities of universal service and equity shown in Fig. 7.1 and the addition of a new priority on electrification for rural development. The universal service priority would remain, albeit with an explicit recognition that it is only one priority among many, that households have differing needs, and that the high costs of household electrification have to be met in a sustainable manner.

The urban/rural price equity priority would be changed to one of keeping electrification affordable, but reflective of costs, willingness to pay and ability to pay for differing levels of service. This does not mean the complete elimination of subsidies. Many successful models involve some subsidies, but they also have to cover some capital costs and pay O&M or they collapse. Models that rely too heavily on subsidies for continued operation are only sustainable as long as the political will to maintain subsidies exists. A change in priorities or economic changes that reduce the ability of either the government or urban consumers to pay the costs of rural electrification threaten any model that is reliant on high subsidies. Replicability is also difficult when capital and operating subsidies are too high. The following chapter discusses the issue of financing distributed electrification and options for reducing prices to rural customers further.

The third priority that has to be made explicit is to link rural electrification more closely with rural development. This would shift the focus off minimal household electrification and on to a more comprehensive approach to electrification that includes productive activities and other welfare enhancing uses of electricity. Ideally, the linkages between electrification and other policy priorities (such as health, education and poverty alleviation) are explicitly recognized and local solutions that address these needs are encouraged and supported. By taking a more comprehensive approach, the revenues from electrification can be increased (due to the higher ability and willingness to pay of productive users) and the effectiveness of investments in electrification should be higher due to the linkage with welfare enhancement goals. The new mix of customers and demand allows for natural market segmentation, improving the viability of distributed electrification efforts.

These shifts in the policy priorities are shown in Fig. 7.2. Rather than being forced towards a centralized solution, this mix of policy priorities can lead to a mix of decentralized generation solutions. However, as made clear by the discussion of local scale solutions above, a key component has to be local implementation and involvement in decision making processes. This can be completely separate from any governmental actions (i.e. local entrepreneurs), in which case the key is not interfering in ways that destroys markets. It can also be the result of government initiatives and partnerships that preserve local market segmentation opportunities and flexibility.

Reconceptualizing Infrastructure Decentralization

The two infrastructure characteristics generally considered important for decentralization are that the service benefits are local and that economies of scale are limited (Estache and Sinha 1995). This has generally limited consideration to infrastructures such as water supply, transportation and waste. However, it is increasingly being recognized that the electric power sector is not limited to

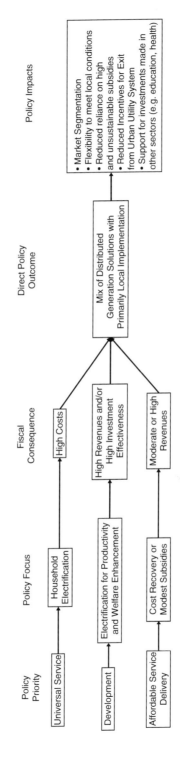

Fig. 7.2 Impact of revised policy priorities

large-scale centralized systems and could be a candidate for decentralization (Estache and Sinha 1995; Bardhan 2002). A few key points can be noted that may be relevant for understanding its impacts on distributed rural electrification.[10]

Decentralization of infrastructure provision is generally considered to have a greater chance for success, when three conditions can be met:

[1] that the local decision process is fully democratic in the sense that the costs and benefits of decisions are transparent and that everyone affected has an equal opportunity to influence the decision;

[2] that the costs of local decisions are fully borne by those who make the decisions, i.e., there is no "tax exporting" and no funding at the margin from transfers from other levels of government; and

[3] that the benefits (like the costs) do not "spill over" jurisdictional boundaries (Bird 1994).

These conditions are considered necessary to balance the decentralized budget (both revenue collection and expenditures must be decentralized in tandem in order to avoid creating imbalances between local and central budgets).[11]

In the case of distributed rural electrification, condition 3 is clearly met since the electricity is produced and consumed locally within the jurisdiction and the consumers of the electricity are the ones who benefit. Conditions 1 and 2 may be more difficult to meet however. Whether condition 1 will be met or not is dependent upon the nature of the local political-economic structure of the community, as discussed above. Given the realities of rural electrification, whether through grid extension or distributed systems, there are strong arguments to be made for continued central financial support even if the decision-making and implementation is decentralized to lower levels of government. These arguments are largely based on the idea that resources (financial, technical, and otherwise) are not available at the local level and require some input from the center (Foley 1992). Furthermore, some degree of central government involvement may be necessary in order to meet redistributive and equity goals (Shin 2001). The question then becomes how to manage and organize that support in ways that do not undermine the rural electrification goals. Also implicit in these conditions is a consideration only of decentralization of government functions. There is an open question as to what the relationship is between decentralized infrastructure provision by private actors and political decentralization of governmental infrastructure decision-making. China provides examples of both how to do this and how to not do this effectively. Their support of the small hydropower and Inner Mongolian wind/pv industries show that strong central support through technology development assistance and financial leverage can result in sustainable and widespread diffusion. On the other hand, top–down led decentralization such as in

---

[10]It should be noted that much of the early work on decentralization that developed the understanding of these advantages were based on industrial country contexts where local institutions are stronger and regulative mechanisms such as elections ("voice") and movement to other jurisdictions ("exit") are much easier (Litvack et al. 1998).

[11]Bird does note that even if these conditions are satisfied, projects may not be successful due to the possible lack of administrative or technical capacity at the local level.

the Township Electrification Program has been successful in the short term but raises significant long-term sustainability questions.

Decentralization of electricity provision can also have implications for the local political economy and result in outcomes that are sub-optimal. Electricity can be seen as a status symbol; if the distributed system is only available to the wealthier households within the community, this can reinforce local elites. Depending on the degree of universal access to electricity and its uses, electrification can therefore serve to either exacerbate or reduce local economic disparities. An important aspect of understanding the local environment for small-scale power is to understand who pays and who benefits from the technology. Technologies and distributed electrification models can be distinguished according to whether they result in private or public benefits and costs. These benefits and costs can be assessed in financial, political and economic terms (Agarwal 1986). A number of different combinations of costs and benefits can apply in rural distributed power depending on the application and the organizational model. Self-generation by rural commercial enterprises incurs private financial costs and returns private financial benefits. On the other hand, a community micro-grid would incur public and private financial costs, but could result in both private financial benefits (to a commercial enterprise, for example) and public economic benefits (through improved quality of life and income earning possibilities due to higher education and better health).

Decentralization of financing can also lead to capture of local government by elites and lead to results that are undesirable from a social welfare viewpoint (Bardhan and Mookherjee 2006). It should be noted that Bardhan and Mookherjee's theoretical analysis was in the context of an infrastructure that is produced centrally, but whose provision could be controlled by either the central or local government (thus, their theory needs amending to account for local production, which is beyond the scope of this chapter). Furthermore, their study compared three modes of financing: local revenue generation through taxation, user fees and central grants. Their results indicate that user fees are generally preferable for both service delivery and welfare. However, in their model, users are assumed to be able to pay for service. This is counter to the assumption usually driving rural electrification efforts. However, as this study shows, Bardhan and Mookherjee's assumption may very well be the one that is closer to the truth, even at low income levels. This does mean that protection from predatory monopoly pricing and capture of electrification benefits from local elites must be part of any regulatory structure.

With these cautionary notes in mind, the evidence from the cases examined for this study shows that decentralization of efforts results in electrification that is more likely to be sustainable and replicable while also contributing to the actual needs of the communities. Global electrification goals are important and central actors need to participate in the electrification process in order to ensure that the resources are available and the institutional incentives are aligned for electrification to spread. However, implementation of electrification activities benefits from having local actors involved or even in the lead.

Flexible and Appropriate Regulations

One consequence of a having an industry dominated by large centralized technologies is that the regulatory system is designed around those technologies. This can create numerous problems for distributed power systems across all sectors and contexts. Licensing and technical requirements designed for larger power plants can create onerous requirements when applied to smaller facilities, resulting in costs that make installation uneconomic (ESMAP 2001). Another major issue is the creation of exclusive service territories for utilities. This can prevent multi-user installations and the creation of local micro-grids because these would, in effect, form small utilities within the exclusive monopoly territory of the larger utility (ESMAP 2001; Morgan and Zerriffi 2002).

Achieving universal service in a way that minimizes hardship on poorer populations while still allowing a variety of approaches will require not only innovative policies but also changes in many regulatory systems. Current regulations often act at cross-purposes to the efficient delivery of services through distributed options. This is the result of both regulatory structure and regulatory practice. Regulatory systems have generally been put in place with centralized utilities in mind. It is difficult for such centralized regulatory systems to effectively monitor a large number of smaller operators. Similarly, the regulatory burdens on the smaller operators of a system designed for large utilities can be prohibitive. For example, licensing procedures can be quite complex and expensive and technical standards for centralized grids may be too strict for local mini-grids or off-grid options and mandate service quality in excess of local needs. These regulatory problems can be handled in a variety of ways, including the creation of standardized licences and flexible power quality and reliability standards.

A significant step forward in developing new regulations that are appropriate for distributed electrification was made with the publication of a paper by the World Bank (Reiche et al. 2006). In the paper, Reiche et al. lay out four principles for regulating electrification:

1. Light-Handed and Simplified Regulation: Complying with regulatory requirements can be time-consuming and expensive. In the case of distributed electrification systems, complicated regulations can also be unnecessary and particularly burdensome for enterprises that are small and barely profitable. Creating a regulatory system that is highly simplified and that does not impose requirements unsuitable for the context will help in creating a legal framework that is broad enough to include smaller actors.
2. Delegate or Contract Out Regulation: Many countries have two bodies, one which is responsible for regulating the power sector and the other responsible for administering government programs on rural electrification. The latter may be better suited to performing the functions of the regulator in rural areas since it will have more extensive information and should be able to better understand how different regulatory requirements will affect rural electrification efforts.
3. Vary Regulation by Type of Entity: There is a diversity of actors that exist in the rural electrification sphere and they are an important part of any decentralized

rural electrification effort (as articulated in the alternative vision). Such diversity requires flexibility in the regulations in order to account for the particular concerns raised by different types of distributed electrification models.

4. Establish Realistic and Affordable Quality Standards: Regulating standards of quality is a thorny problem. On the one hand, there are few quality standards in existence today, leading to poor service. On the other hand, the level of quality needed may be quite different and context dependent and imposing uniform standards of quality can result in too rigid a system. For example, households at the lowest end of the income scale may not need or be willing to pay for electricity that is available 24/7 except for some key applications (e.g. health clinic vaccine refrigeration). Flexibility needs to be built into the standards and into the enforcement of standards.

Most importantly, the authors go on to describe the standards for a model law on electrification that would go a long way towards resolving the problems with current electrification programs. Many of the barriers to entry that currently exclude viable decentralized solutions would be removed by the proposed law and regulatory principles.

However, the model law may not help solve some of the basic problems with current rural electrification programs. It still allows for market distorting subsidies and has improving the number of connections as, at least implicitly, a major determinant of performance. Absent a corresponding set of principles to govern subsidies and financing of electrification and the incorporation of an overarching development priority for electrification, the centralized model based on high subsidies and minimal household electrification could continue.

Restructuring the Social Contract

The restructuring of electric power systems worldwide has raised concerns that the reform process would undue the "social contract" that had been struck between governments, the electricity industry and the population at large. In most places the social contract was an exchange in which the government regulates the industry and guarantees its financial viability while ensuring protection of the poor and the environment (which are not easily integrated into a pure market) (Heller et al. 2003; Chaurey et al. 2004). The types of changes discussed in this book are similar to some of the changes undertaken in the larger market reform processes. There is a greater role to be played by the private sector and an emphasis placed on rationalizing the finances of rural electrification efforts. It also calls for a move away from highly centralized electrification processes and towards a more decentralized and heterogeneous approach.

So, while this vision of distributed electrification is based on the search for solutions on a local scale, this does not absolve centralized governments of their responsibilities nor does it call for a complete removal of the international donor community from solving the problem. However, it does envision a different role for both higher level government agencies and international donors (both official donor aid and NGO aid). This is necessary in order to maintain the social contract that previously existed (though perhaps in a different form). Just as a system completely dominated by centralized decision-making seems to have problems in creating a viable, sustainable and replicable model for distributed electrification, those that arise from a lack of centralized action are usually serving a small niche of the market and often with low-quality and expensive power. The key is to find a role for non-local actors to meet certain societal goals but that does not destroy the market segmentation, local needs-based decision making and possible contributions to greater rural development that comes with smaller scale solutions. Although it is focused on solar systems, the following list of recommendations is a good starting point for understanding the role of governments in supporting the creation of viable rural electrification models:

(i) supporting and conducting least-cost rural energy planning that includes PV system options;

(ii) making investment capital available for SHS programs;

(iii) encouraging the commercial banking sector and financing agencies to finance PV home systems on reasonable terms by offering support mechanisms such as refinancing arrangements;

(iv) supporting promotional campaigns for PV household systems among rural households;

(v) removing regulatory barriers that limit competition among energy service providers;

(vi) offering training and technical assistance to help establish retail and service networks. (Cabraal et al. 1998)

All of this would be greatly aided if the various programs that are geared towards rural electrification could be performance based while at the same time reducing the costs associated with coordination. In many cases, multiple ministries might be involved (e.g. those in charge of energy, agriculture, development, finance) or may even have duplicative or competitive programs. For example, in India there is a Ministry of Power, but also a Ministry for Non-Conventional Energy Sources, with obvious overlaps (Radulovic 2005; Srivastava and Rehman 2006). Lack of coordination increases transaction costs because those seeking either approval or funding may have to deal with multiple ministries to determine which one is appropriate (or more accurately, which *ones* are appropriate). Another area in which coordination may be useful is between energy programs and other development focused programs for which energy may be a useful input (Radulovic 2005; Srivastava and Rehman 2006). For example, rural health and education are far too often treated as entirely separate from the problem of rural energy. Closer ties between programs would help in ensuring that the necessary infrastructure is being developed in conjunction with expansion of social programs.

One of the major impediments to distributed rural electrification is the lack of local organizational, financial and technical capacity to implement distributed rural electrification. Decentralization of decision-making on infrastructure without strengthening the necessary local institutions (either private or public) will decrease the likelihood of success (Bird 1994). Many of these types of challenges to implementing a more decentralized vision of rural electrification can also be reduced through concerted activities of the international donor community (including the multilaterals, bilateral donors and NGOs). Financial support will remain an important part of their programs, particularly aid to help in reducing the first-cost problem through small-scale lending or other innovative financial mechanisms. Distributed efforts would also be helped by the type of larger programmatic activities that can only be engaged in by more centralized actors, including aid for reforming regulations and developing sound policies. The donor community can also complement its financial support with capacity building in areas such as business management and technology maintenance.

Conclusion

In the literature on decentralized power generation, this push towards local generation is usually coupled with technology solutions that are more environmentally friendly. The lack of existing infrastructure provides an opportunity for a technological leap-frog over the dirty technologies of the industrialized countries and the creation of energy paths that are clean and sustainable. The renewable "small is beautiful" image is certainly compelling. However, unlike the image often portrayed, the technologies that are actually used range quite widely and include fossil-fuel based options such as diesel generators. There is also a stunning diversity in business models and diffusion methods that have been used to provide these technologies in rural areas. These options, just as with grid extension into rural areas, are not cheap and the choices made in terms of capital financing, recovery of operations and maintenance, target customers and the type of organizations involved all matter to the success and failure of these ventures.

This creates a much more complicated image, or really sets of images, of distributed rural electrification. The face of rural electrification is the well-managed centralized utility distribution system and the farmer with her functional solar home system. But it is also failing utilities unable to meet demand and solar home system programs that install systems that no longer work after a couple of years. It is also diesel generators and small hydropower and it is a whole host of actors ranging from international donors and national ministries to entrepreneurs, NGOs, cooperatives and local governments. These different models fit into the twin goals of universal service and low prices in different ways. They also have different implications for economic development and improved social welfare.

References

Agarwal B (1986) Cold hearths and barren slopes: the woodfuel crisis in the third world. Zed Books Ltd, London

Allderdice A, Rogers JH (2000) Renewable energy for microenterprise. National Renewable Energy Laboratory, Golden, CO, pp 1–70

Bardhan P (2002) Decentralization of governance and development. J Econ Perspect 16(4):185–206

Bardhan PP, Mookherjee PD (2006) Decentralisation and accountability in infrastructure delivery in developing countries. Econ J 116(508):101

Barnes DF (1988) Electric power for rural growth: how electricity affects rural life in developing countries. Westview Press, Boulder, CO

Barnes DF, Floor WM (1996) Rural energy in developing countries: a challenge for economic development. Annu Rev Energy Env 21:497–530

Barnes DF, Halpern J (2000) Subsidies and sustainable rural energy services: can we create incentives without distorting markets? Joint UNDP/World Bank Energy Sector Management Assistance Programme (ESMAP), Washington, DC, p 13

Barnett A (1990) The diffusion of energy technology in the rural areas of developing countries: a synthesis of recent experience. World Dev 18(4):539–553

Bird R (1994) Decentralizing infrastructure: for good or ill? The World Bank, Washington, DC, p 34

Bond TC, Sun H (2005) Can reducing black carbon emissions counteract global warming? Environ Sci Technol 39(16):5921–5926

Cabraal A, Davies-Cosgrave M et al. (1998) Accelerating sustainable photovoltaic market development. Prog Photovoltaics Res Appl 6:297–306

Cabraal RA, Barnes DF et al. (2005) Productive uses of energy for rural development. Ann Rev Environ Resour 30:117–144

Chakravarty S, Chikkatur A et al. (2009) Sharing global CO_2 emission reductions among one billion high emitters. Proc Natl Acad Sci 106(29):11884–11888

Chaurey A, Ranganathan M et al. (2004) Electricity access for geographically disadvantaged rural communities: technology and policy insights. Energy Policy 32(15):1693

Cherni JA, Dyner I et al. (2007) Energy supply for sustainable rural livelihoods. A multi-criteria decision-support system. Energy Policy 35(3):1493–1504

Institui o Programa Nacional de Universalização do Acesso e Uso da Energia Elétrica (2003) "LUZ PARA TODOS" e dá outras providências. Brazil. Decreto No – 4.873.

Dossani R (2004) Reorganization of the power distribution sector in India. Energy Policy 32:1277–1289

Easterly W (2006) The White man's burden: why the west's efforts to aid the rest have done so much ill and so little good. Penguin Books, New York

ESMAP (2001) Best practice manual: promoting decentralized electrification investment, Energy Sector Management Assistance Programme

Estache A, Sinha S (1995) Does decentralization increase spending on public infrastructure? The World Bank, Washington, DC, p 26

FAO (2000) The energy and agriculture nexus. Food and Agriculture Organization of the United Nations, Rome, p 84

Fishbein RE (2003) Survey of productive uses of electricity in rural areas. Africa Energy Unit, World Bank, Washington, DC, pp 1–50

Foley G (1992) Rural electrification: the institutional dimension. Utilities Policy 2(4):283–289

Goldemberg J, Johansson TB (eds) (1995) Energy as an instrument for socio-economic development. UNDP/BDP Energy and Environment Group, New York

Government of Cambodia (2004) Percentage of households owning consumer durable goods, Cambodia Socio-Economic Survey. National Institute of Statistics, Government of Cambodia, Phnom Penh

Government of India (1992) The Constitution (Seventy-Third Amendment) Act

Government of India (2006) Integrated energy policy: report of the expert committee. Planning Commission, Government of India, New Delhi, p 182

Hart SL (2005) Capitalism at the crossroads: the unlimited business opportunities in solving the world's most difficult problems. Wharton School Publishing, Upper Saddle River, NJ

Heller TC, Tijohn HI et al. (2003) Electricity restructuring and the social contract. Program on Energy and Sustainable Development, Stanford, CA

Int.Don5 (January 2006) Author interview with international donor. Washington, DC

International Energy Agency (2006) Energy for cooking in developing countries. World energy outlook. International Energy Agency, Organization for Economic Co-operation and Development, Paris, pp 419–446

Litvack J, Ahmad J et al. (1998) Rethinking decentralization in developing countries, World Bank, p 52

Martinot E, Chaurey A et al. (2002) Renewable energy markets in developing countries. Annu Rev Energy Env 27:309–348

Meadows K, Riley C et al. (2003) A literature review into the linkages between modern energy and micro-enterprise. AEA Technology plc for the UK Department for International Development, Oxfordshire, p 33

Morgan MG and Zerriffi H (2002) The regulatory environment for small independent micro-grid companies. The Electricity Journal 52–57

Nieuwenhout FDJ, van Dijk A et al. (2001) Experience with solar home systems in developing countries: a review. Prog Photovoltaics Res Appl 9:455–474

Pan J, Peng W et al. (2006) Rural electrification in China 1950-2005. Research Centre for Sustainable Development, Chinese Academy of Social Sciences and Program on Energy and Sustainable Development, Stanford University, Beijing, China and Stanford, CA

Prahalad CK (2005) The fortune at the bottom of the pyramid. Wharton School Publishing, Upper Saddle River, NJ

Radulovic V (2005) Are new institutional economics enough? promoting photovoltaics in India's agricultural sector. Energy Policy 33:1883–1899

Ramani KV, Heijndermans E (2003) Energy, poverty and gender: a synthesis. International Bank for Reconstruction and Development, Washington, DC

Reiche K, Tenenbaum B et al. (2006) Promoting electrification: regulatory principles and a model law. Energy and Mining Sector Board, World Bank, Washington, DC, p 49

REN21 (2008) Renewables 2007 Global Status Report. REN21 Secretariat and Worldwatch Institute, Paris and Washington, DC, p 51

Shin R (2001) Strategies for economic development under decentralization: a transformation of the political economy. Int J Public Admin 24(10):1083–1102

Srivastava L, Rehman IH (2006) Energy for sustainable development in India: Linkages and strategic direction. Energy Policy 34:643–654

Stewart F (ed) (1987a) Macro-policies for appropriate technologies in developing countries. Westview Press, Boulder, CO

Stewart F (1987) Macro-policies for appropriate technology: an introductory classification. Macro-policies for appropriate technology in developing countries. Westview Press, Boulder, CO, pp 1–21.

Thakur T, Deshmukh SG et al. (2005) Impact assessment of the Electricity Act 2003 on the Indian power sector. Energy Policy 33:1187–1198

Tong J (2004) Small Hydro Power: China's Practice. China WaterPower Press, Beijing

Tongia R (2007) The political economy of Indian power sector reforms in Victor D and Heller T eds. The Political Economy of Power Sector Reform The Experiences of Five Major Developing Countries. (Cambridge: Cambridge University Press): pp 109–174

van Campen B, Guidi D et al. (2000) Solar photovoltaics for sustainable agriculture and rural development. Food and Agriculture Organization, United Nations, Rome, pp 1–76

World Bank (2003) Project appraisal document on a proposed credit in the amount of SDR 27.9 million (US$40 million equivalent) and a proposed GEF grant of US$5.75 million to the Kingdom of Cambodia for the rural electrification and transmission project. Washington, International Bank for Reconstruction and Development, Energy and Mining Sector Unit, SE Asia and Mongolia Country Unit, East Asia and Pacific Region, p 146

World Resources Institute (2009) Climate analysis indicator tool

Zerriffi H, Wilson E (2010) Leapfrogging over development? Promoting rural renewables for climate change mitigation. Energy Policy 38(4):1689–1700

Chapter 8
Paying for the Vision: New Financial Models for Distributed Electrification

Keywords Donors • Financing • Governments • Institutional reform • Microfinance • Private sector

Introduction

As noted throughout this book, rural electrification is a challenge because the costs of service are generally high and the willingness and ability to pay of the majority of customers are generally low. The prior chapter outlined a new vision for rural electrification based on policies that would include rural development as a priority and would emphasize affordability of rural electrification costs. This would require a shift in policies and, in many places such as Brazil, a shift in the regulatory system to remove technical barriers. However, even with those changes, the underlying financial difficulties inherent in rural electrification would remain. In particular, there would still be legitimate issues with the ability of rural populations to pay for electricity and equity issues regarding services for urban and rural consumers. This chapter addresses these issues directly.

While many people in the world live without electricity, others get their electricity for free or at a price that is well below the cost of service. Those costs are then borne directly and indirectly by other segments of society. In some cases, it is the urban customers who pay directly through cross-subsidization. In other cases, the economic costs are spread throughout society in the form of budgetary shortfalls, poorly performing utilities or higher prices.

There are three important challenges to the financial viability of distributed electrification and they are covered in the next section. The most obvious financial issue is the technical cost of distributed electrification, absent any subsidy structure and how it compares to the costs of traditional grid extension. With those costs in mind, the next sub-section section addresses the problem of the willingness and ability to pay of rural customers. The examples from the cases, particularly Cambodia, show that willingness and ability to pay may be higher than usually assumed. The final challenge is access to financial resources on the local level, particularly for covering capital costs. The usual solution is to create large-scale subsidies for rural electrification. The role of different types of subsidies and the downsides to subsidies is the topic of the third section. The final section sets out some guidelines for rural electrification policies that seek to address some of the hurdles outlined in the previous sections and create the conditions for implementing sustainable and replicable models for distributed electrification.

Financial Challenges to Distributed Electrification

Technical Costs

While there are a variety of factors that influence whether a distributed technology gets installed and used (as opposed to either no electricity or the grid system being extended), one of the key factors is the technical cost of the system. There are a wide variety of technological options for providing rural electricity using distributed power generation. A summary of key characteristics of distributed generation technologies suitable for rural electricity supply is given in Table 8.1. Included are current and future costs as well as the suitability of the technology for either individual installations (either in homes, community buildings or for productive activities) or in microgrids. By comparison, one estimate puts the generation costs of baseload power plants in the centralized grid system as low as 5–7 cents/kWh plus another 3 cents/kWh for transmission (costs unadjusted for inflation) (World Bank 1996). However, when factoring in transmission and distribution to rural areas, the costs can rise significantly. Depending on the density of households and length of the grid extension, another 2–45 cents/kWh can be added on to the generation and transmission costs (costs unadjusted for inflation) (World Bank 1996). The particular numbers are not as important as the fact that trying to serve low density communities through the grid can be significantly more expensive than serving urban consumers, with grid extension and distribution costs far outweighing costs of generation.

Even between rural options, there are a number of choices that will depend on a variety of factors including the density of consumers and the number of households served. For example, Cabraal et al. (1998) plot the break-even line between PV and a local grid in an isolated village based on data from Indonesia. This is based on an "Equivalency of Service" concept such that the energy services used by the different households are the same. In general, their findings show photovoltaics to be the lowest cost option for low density and low total population communities, particularly those farther from existing low and medium voltage lines. In their particular example, photovoltaics make the most financial sense in isolated villages anytime there are less than 60 households per square kilometer or less than 200 households in total (Cabraal et al. 1998). They also plot data for villages located 5 km from a medium voltage line and 3 km from a low voltage line.

Table 8.1 Technical characteristics and potentials for cost reductions for select distributed generation technologies (Sources: Dunn (2002) for size, efficiency and capital costs for engines, wind turbines, and solar. REN21 Renewable Energy Policy Network (2005) for wind, biomass gasification, small hydro capital cost and size data and all levelized costs and total generation figures. Banerjee (2006) for diesel generation costs)

Technology	Reciprocating engine	Wind turbine	Solar	Biomass gasification	Small hydro[a]
Size (kW)	5–10,000	<1–3,000	<1–1,000	20–5,000	<1–50,000
Efficiency (%)	20–45	–	–	–	–
Current capital cost ($/kW)	600–1,000	850–1,700	5,000–10,000	500–4,000	1,000–5,000
Levelized cost (cents/kWh)[b]	~10[c]	Large (1–3 MW): 4–6 Small (3–100 kW):15–30 Micro (0.1–1 kW): 20–40	Large (2–5 kW): 20–40 Small (20–100 W): 40–60	8–12	Small (1–10 MW): 4–7 Mini (0.1–1 MW): 5–10 Micro (1–100 kW): 7–20 Pico (0.1–1 kW): 20–40
Future capital cost ($/kW)	<500	500	1,000–2,000		
Total installed generation	N/A[d]	N/A[e]	2 million systems[f]	N/A[g]	61 GW
Operating modes	Individual installations (for productive activities), microgrids	Individual installations, microgrids (often as wind-diesel hybrids)	Individual installations, microgrids (often as solar-diesel hybrids)	Individual installations (for productive activities), microgrids	Individual installations, microgrids

[a]The definition of small hydro varies from country to country, reaching as high as 50 MW in China. More commonly the cutoff is considered to be somewhere in the range of 10–25 MW. Total installed generation is based on each country's definition of small hydro (REN21 Renewable Energy Policy Network 2005) and so could include 50 MW plants from China, but exclude 50 MW from countries with a lower threshold

[b]Levelized costs include both capital costs and costs for operations and maintenance. Capital costs are amortized, making assumptions regarding discount rates and operating lifetimes. Operating costs depend greatly upon resource availability, capacity factors and host of other site specific and operation specific parameters and assumptions must be made here as well. Levelized costs are from (REN21 Renewable Energy Policy Network 2005) and calculated "under best conditions, including system design, siting, and resource availability." As such, these cost ranges can be considered typical but may not encompass the full range of levelized costs, particularly on the higher end (if facilities are run particularly poorly or in very difficult conditions costs could rise significantly). The specific assumptions, including cost of end-use demand, were not provided, so it was not possible to judge the basis for the estimates. However, they are in-line with other estimates for renewable energy costs

(continued)

Table 8.1 (continued)

^cUnlike renewable energy technologies such as solar and wind, fuel purchases are a very important component of the cost, making the levelized cost highly variable over time as a result of fuel supply price changes. As a result of these fuel costs, high losses, poor maintenance and a variety of other reasons, it is not uncommon to have high prices for diesel engine based micro-grid electricity (~20–60 c/kWh). (World Bank 1996) For example, in Cambodia prices can easily reach ~50 cents/kWh. The big difference is that capital costs are much lower than the renewable sources, making initial installation more viable and then it is possible, at least in principle, to keep costs below the minimum costs of the alternative renewable technologies in areas where diesel fuel supply is not a problem

^dIt is well known that diesel engines (and gasoline to lesser degree) are widely used in both urban and rural settings. While they are acting as back-up generators in some cases, there is a widespread use of engines as the sole electricity source. However, unlike with renewables, there is no coordinated effort to collect data on their use

^eData on overall wind turbine installations is available, but it is difficult to find data on the small wind turbines that would be most suitable for local generation in rural areas

^fData on the small, off-grid PV systems that are often used in rural areas is often given as the number of units sold or installed. The size of individual units can vary widely making it difficult to determine exactly how many MW have been installed. It should also be noted that the Township Electrification Program of the Chinese Government installed roughly 20 MW of solar in between 2001 and 2005

^gFigures for small community level biomass gasification is hard to come by as it remains a technology that is installed for demonstration purposes but has not expanded to widespread application (as opposed to large MW scale biomass facilities and other ways of using biomass such as direct burning or landfill methane burning)

Willingness and Ability to Pay

While rural communities will generally be at the bottom of the socio-economic ladder within a country, there exists great variation among rural communities and even within a given community in terms of socio-economic status. This directly affects the ability to pay for electricity service and thus the amount of electricity service that will be demanded (both directly and implicitly through the ability to pay for electricity consuming devices). Low ability and willingness to pay is often cited as a reason for large and subsidized rural electrification programs.

While the literature and data exploring these relationships is far from complete, available data indicates that primary energy consumption may take on a U-shaped functional form with respect to income. At the lowest income levels, very poor quality fuels (such as firewood) are being used and overall energy consumption is high. Much of this consumption may be in the form of collected fuelwood and therefore outside the market system if self-collected. At the same time, the proportion of household income spent on energy can be highest at the lowest end of the income scale. Households at slightly higher income levels may be able to take advantage of modern energy services, reducing overall energy consumption, but may not have the disposable income to purchase high energy consuming devices. At the highest end of the income scale, households are able to purchase devices such as refrigerators that consume more electricity (as well as non-electric high energy consuming devices such as mopeds, increasing their overall energy consumption) (ESMAP 2000; van Campen et al. 2000; Elias and Victor 2005).

In Brazil, the official prices have been kept artificially low and quantity has been controlled as well. For example, COELBA only provides solar home systems of a single size and the diesel generators run by CEAM are of limited capacity. This makes the willingness or ability to pay difficult to judge. However, the presence of alternative models in which prices are higher (e.g. IDEAAS, BRASUS and independent diesel micro-grids) demonstrates a generally higher willingness to pay and a higher ability to pay among some segments of the rural Brazilian population. Survey data from three Brazilian states also shows significant diversity in willingness to pay (ESMAP 2000).

The Cambodian case demonstrates than in some places there can be a high willingness to pay across the rural population. The high prices in Cambodia do result in low consumption, indicating that ability to pay is relatively low, even if the willingness is high. Of course, as with the Brazilian case, there are subsets of the population whose ability to pay is higher (the solar home system customers and some customers within the diesel micro-grids).

Finally, in China the evidence is also strong that willingness to pay is generally high and ability to pay among a subset of consumers is equally high. The evidence for the willingness to pay comes from the markets for renewables in the northwest and the Inner Mongolia Autonomous Region. The differences in the size of the units sold in those markets shows that there is a diversity in ability to pay. The units sold in the IMAR are larger, reflecting the better economic conditions for many rural consumers in that region due to their high value product. In the northwest, the small size of the solar home systems is the direct result of a low ability to pay among the consumers.

Willingness to pay will depend on a variety of factors, including income level, the price of substitute fuels, the level of electrification and real or perceived availability of alternatives for electrification. One important factor is the existing energy budget of the household. Many households without access to electricity still spend a significant portion of their household budget on energy services. Lighting by kerosene, candles, or rechargeable car batteries can be extremely expensive for poor quality service. Evidence indicates that these households are often willing to pay high prices per kWh for electricity in order to obtain better service and energy quality (van Campen et al. 2000; ESMAP 2001; Illskog et al. 2005). However, expectations that electricity service (particularly grid-based service) will bring parity with urban consumers can result in pressure for a uniform tariff and reduce the willingness to pay of rural consumers. Another consideration regarding willingness to pay is the quality of the service provided. It is commonly asserted that Indian farmers who currently pay nothing for very poor quality power (so poor that it sometimes even harms their water pumps) would actually be willing to pay for better quality service (ESMAP 2002).

While willingness to pay and ability to pay are related, they must be distinguished in considering the nature of the customer base. The customer base for a distributed rural electrification project might include households of differing economic status, community buildings, and productive enterprises. The ability to pay of these three customer classes will be different and will affect the underlying economic viability of any organizational or business model that seeks to meet the needs of these customers. In some cases, measuring willingness to pay through survey methods can help determine how best to meet the overall demand. However, care must be taken when using willingness to pay from surveys as the willingness to pay of some customers may be higher than their actual ability to pay and their demand may change once they are billed.

Access to Financial Resources

One of the major institutional barriers to decentralized power systems is the lack of access to capital by both end-use customers (households and commercial enterprise) and by decentralized commercial entrepreneurs. The obvious solution would be to provide access to credit for capital purchases, yet the problems of providing a large number of small loans to individuals with little credit history are well known and commercial lenders (or even state-owned banks) are reluctant to enter into such loans. In some cases, there are financial institutions that are geared towards low-income rural borrowers, such as agricultural banks. However, these banks have a difficult time assessing the risks of these loans. The technology and/or the business model may be unfamiliar to them and make them reluctant to enter this lending market (Martinot et al. 2002; Radulovic 2005).

As discussed in Chapter 3, the Brazilian Luz Para Todos program has significantly expanded the financial resources available to the utilities for meeting their rural electrification commitments. Utilities can also count on the availability of commercial loans and the BNDES to finance projects as well as the cross-subsidization of rural customers by urban customers. By contrast, access to financial resources is more difficult for alternative actors in Brazil. They must rely on funds from international donors and NGOs. Access to private equity via formal mechanisms is more difficult.

In Cambodia there is little official money for electrification projects in rural areas. The central government's primary goals have more to do with importing power from neighboring countries and helping develop larger power plants to meet the needs of the urban and industrial sectors. Some money has come into the country from NGOs and foreign donors, but not at a significant rate. The rural electricity entrepreneurs appear to rely primarily on informal lending networks which result in extremely high cost of capital.

One way in which the Chinese system shows a high degree of centralized planning towards a single goal (universal electrification) is in its lending institutions. The central banks that operate in rural China have a mandate to lend for infrastructure projects. Interest rates for small hydro projects are often close to zero and loans are never defaulted even for lack of payment. This is, perhaps, reflective of other systemic problems in the Chinese banking sector and such treatment may not be as readily available in the future.

The Conventional Solution: Subsidies

The conventional solution to the problem of high costs and low willingness (or ability) to pay is to reduce either the capital costs or the consumption costs of electricity to the end-user through some form of subsidy, while leaving the rest of the institutional structure largely intact. Subsidies are more generally justified as a means to correct for market failures. In this case, the market may not be able to provide electricity in rural areas in a way that meets social expectations and equity criteria. Like public goods, which are often poorly supplied by markets, these are social goods and the overall welfare impacts of using subsidies to provide those goods may be justified (Barnes and Halpern 2000; UNEP 2002).[1]

However, like all subsidies, these are distortions to the market, and their net effect can be positive or negative depending upon the intended goals of the subsidy and the details of the subsidy implementation. A key part of achieving more widespread electrification will be addressing the issues regarding willingness/ability to pay and equity issues, particularly urban vs. rural and this will likely result in some form of subsidies. However, if subsidies are to be used to implement a new vision of rural electrification, it is necessary to understand the various shortcomings of the subsidy system and develop guidelines for effective subsidies. This section reviews a number of the key issues with subsidies while the next section includes a discussion of the reforms needed for subsidies to contribute to achieving the new vision for distributed electrification.

---

[1] For a standard introductory textbook explanation of the limits of markets, see Nicholson (1998).

Types of Subsidies

A comprehensive review of the field of subsidies is beyond the scope of this book. However, there are key elements of subsidy systems that are worth reviewing briefly. The first distinction that can be made is between subsidies for capital versus subsidies for consumption. Capital cost subsidies can come in various forms: technology dumps (i.e. giving away solar home systems for free), low interest loans, grants, payments as part of technology programs. They are all geared towards reducing the cost of capital investments either by the end consumer or by a service or technology provider. Consumption subsidies, on the other hand, are designed to reduce the regular costs of obtaining electricity service. This can include free electricity programs or reduced tariffs that are below the cost of service.

Distinctions can also be made according to how the subsidy is implemented. Subsidies can be divided generally into demand-side and supply-side subsidies (Barnes and Halpern 2000). Each have their advantages and disadvantages. Demand side subsidies have the advantage of being accessed directly by the subsidy recipients and, in theory, should be sustainable and replicable since they do not disincentive suppliers. However, they can also be difficult to implement and suffer from high transaction costs (Barnes and Halpern 2000). Supply side subsidies have the advantage of simplicity of implementation but the disadvantages of poor targeting and can create market distortions that can actually undermine service delivery (Barnes and Halpern 2000).

A third useful distinction is between sources of revenue to cover the subsidy. In cross-subsidy schemes, the costs are borne directly and explicitly by others within the same system (e.g. urban customers paying a higher price so that rural customers can pay a lower one). In other cases, the cross-subsidy can be spread further and be more implicit (e.g. CEAM's losses being covered by its parent holding company which also owns the generation systems that provide electricity to Brazil's richer urban centers). Subsidies can also be provided by the central government through ministry funds or by international donors either through donations of money or goods.

As discussed in the prior chapter, Brazil has a highly subsidized system in which significant funds are being currently expended by the central government for capital construction costs. Subsidies for consumption are also provided either through internal utility cross-subsidies or by passing losses up to electricity holding company in the case of government utilities. Cambodia by contrast has no subsidy system other than for individual projects as a result of donor assistance (e.g. the Japanese donated hydro-PV project). China has a much more mixed experience with some highly targeted and modest technology related subsidies, preferential banking and tax policies, and some highly subsidized programs (e.g. the Township Electrification Program).

The Downside of Subsidies

If the purpose of subsidies is ostensibly to correct for failures in the market system to provide social goods, then why haven't been more effective in electrifying rural areas? In reality, of course, subsidies are all too often provided for far more parochial interests, whether it is the financial interests of a powerful lobby or the political interest of a party. More importantly, even when subsidies are implemented for their intended purpose, there are a number of potential problems with subsidies that can make them significantly less effective or even counter-productive.

The most obvious effect of a subsidy regime is the loss of economic efficiency. Even if those losses are considered acceptable in order to meet other goals, they must be recognized and possibly addressed in other ways. Depending on the subsidy, economic efficiency losses arise due to a variety of effects, including over-consumption, lowered return on investment by the private sector, physical shortages of energy, increased trade imbalances due to energy imports or reduced energy exports (UNEP 2002). Those efficiency losses may be justified but they must evaluated on a case-by-case basis.

One major problem with subsidies, including in distributed electrification, is the destruction of viable markets for appropriate technologies (Barnes and Floor 1996). In some cases, distributed electrification technologies cannot compete due to the subsidies provided other alternatives such as kerosene or the grid system (Srivastava and Rehman 2006). In other cases, technologies are promoted through a heavily subsidized mechanism, only to see the market disappear once the subsidy is removed because no effort was made to create self-sufficiency (Barnett 1990; Martinot et al. 2002). In other cases, it is not the technologies that are necessarily adversely affected, but the mechanisms by which those technologies are diffused into rural areas. The subsidies to COELBA for its solar home program ensure that it is the only actor in Bahia, even if it were to lose its exclusive service territory provisions.

Subsidies can also create significant financial burdens on the government, which affects both the electricity sector and possibly other sectors (Barnes and Floor 1996). In India, power sector subsidies have resulted in significant financial impacts on the State Electricity Boards and, as a result, on the budgets of the individual states (Tongia 2007). The opportunity costs of these subsidies, especially given the poor performance of the power sector in effectively using those subsidies is high. One estimate for Uttar Pradesh is that the 3.7 billion dollars spent on power sector subsidies is equivalent to over 300,000 health clinics or over one million schools (Srivastava and Rehman 2006).

It can also be difficult to target subsidies effectively so that those in greatest need are the ones that benefit the most from the subsidy (Barnes and Halpern 2000). Subsidies are often set on the basis of consumption (as in Brazil) which may or may not be a good proxy for income (and are problematic if the consumption level is set too high). Subsidies might end up benefiting producers who may not pass on those benefits to their lower income consumers. Subsidies can also end up benefiting households at the upper end of the income scale rather than those at the bottom. This can result because of the inability of the poorest households to take advantage of the subsidy (i.e. they may not be able to gain physical access or purchase the end-use devices), their low consumption, the unequal access of households to rationed goods, including electricity, or the ability of certain segments of society to capture the benefits of a subsidy program (Barnes and Halpern 2000; UNEP 2002; Srivastava and Rehman 2006). Since the costs of subsidies are spread, either directly or indirectly, throughout society, the poorest households may end up paying for energy subsidies that do not benefit them (UNEP 2002). They may also feel the effects of expensive but ineffectual subsidies through the reduction in other services due to poor government finances.

One area upon which the literature on subsidies is often silent is the link between subsidy programs and centralization of service delivery. As can be seen from the cases studied for this book (see Chapter 6) there is a strong linkage between highly subsidized electrification programs and centralized business models (either government programs, centralized utility programs or international donor programs). However, as discussed below, there may be ways of restructuring the institutional relationships between rural electrification actors in order to break that link and allow for a much wider diversity of business models to take advantage of well-structured subsidy programs.

Expanding the Financial Options

Alternative Financial Models

There are a number of ways to overcome the problems of cost, affordability, and access to financial resources that do not rely entirely upon subsidies. In the case of stand-alone technologies, the first is to reduce the total amount of capital required. For example, the sizes of PV systems can be reduced to make them more affordable, resulting in systems as small as 10 W. The per kW price is still high, but the smaller size allows households with more limited capital to purchase an initial unit. Households can then purchase additional units according to changes in their income. The smallest systems are sold in purely cash markets in countries like Kenya, Morocco, and China. This dealer model can be quite effective though there have also been problems in being able to expand the service territory and capitalization is always a problem (Cabraal et al. 1998; Barnes and Halpern 2000; Martinot et al. 2002).

Another solution that solves the capital cost problem for the consumers are rental models or fee-for-service models. In the rental model, the consumer pays a monthly fee for use of single facility technology (e.g., a solar home system). Ownership remains with the technology distributor, not the end consumer and a flat monthly fee is paid to the owner for use of the technology. This saves the household from having to raise enough capital to purchase the technology outright and the dealer can presumably improve their buying power and access different credit facilities (Barnes and Floor 1996; Cabraal et al. 1998; Barnes and Halpern 2000). In a variant of the model, ownership is transferred to the household once the system has been paid for through the monthly fee (Cabraal et al. 1998).

In the fee-for-service model, it is the energy, not the technology, which is the commodity. Common examples include micro-grid systems run off of diesel generators, small hydro stations, biomass gasification systems, as well as battery charging stations. These Energy Service Companies (ESCOs) can be run by a local entrepreneur, possibly as an adjunct to an existing productive activity (e.g., an agricultural producer who uses the electricity during the day and can provide electric power to local households in the evening), the local government, a cooperative or an NGO. The capital cost problem still remains for the provider of the service however. This is an area of active institutional experimentation with various approaches for incentivizing existing financial institutions to enter into this market as well as setting up new financial arrangements. While still difficult, it is perhaps an easier challenge than the one of providing capital to households for equipment purchases.

Whether as part of other business activities or as a stand-alone business, the micro-grids are essentially small utilities. These micro-utilities can also be part of a larger utility system. In some countries, concessions systems have been established for rural electrification in which the concessionaire has the option to serve some of its customers by running small systems that are disconnected from its larger distribution network (Barnes and Halpern 2000). This solves the problem of making small loans to unknown and risky customers since the larger utility will have access to capital markets and banks and is likely to be receiving some form of government incentive for providing electricity to these off-grid rural areas. However, in many cases the concessionaires have had a difficult time meeting their obligations in these areas at a cost that is acceptable to them.

For the private sector at the local level, one way to address this deficiency is through what are called Market Facilitation Organizations (MFOs). These are "public-private entities that support the growth of particular markets through a variety of means," (Martinot et al. 2002) ranging from more intangible benefits – such as access to information and networking – to technical support and financing. Until now there have been limited attempts to develop MFOs; those attempts have generally been undertaken by NGOs or through the central government (and its branches at the local level). So far the results seem promising. However, those present for rural energy supply tend to focus on renewables rather than on rural electrification.

Micro-credit

One solution to the rural finance problem that has proven successful in a number of non-energy areas is the presence of micro-credit lending agencies. These micro-credit agencies disburse very small loans, often to groups of borrowers (de Aghion and Morduch 2005). Micro-credit loans are usually for purchases that will generate income and result in re-payment over a short period of time (Martinot et al. 2002). Small DG entrepreneurs would seem to fit this description. However, in some cases the loans required to set-up a DG business far exceed the upper limit of micro-credit lending. Thus, entrepreneurs are caught in the valley between very small micro-credit loans and the larger loan sizes of commercial or central banks. There are a number of other potential problems in using micro-credit programs for distributed generation. For one, micro-credit programs have the same problem of familiarity with these technologies as the larger commercial lenders. Additionally, the micro-credit programs are not usually geared towards solving the consumer credit problem, and purchasing items such as solar home systems may be seen as consumer goods rather than productive goods. This would preclude some business models based on end-use sales from benefiting from micro-credit.

This is not to say that micro-credit cannot be used as the example of Grameen Shakti clearly demonstrates. Grameen Shakti is an off-shoot of the highly successful Grameen micro-lending bank started by Mohamed Younis which provides credit for the purchase of solar home systems in Bangladesh. Keys to their success including linking installations to income generating activity and creating a network of entrepreneurs to install and provide service. The result has been 77,000 solar home systems installed plus micro-utilities and biogas plants. The systems are unsubsidized and affordability is through a loan system (Biswas et al. 2004; Uddin et al. 2006).

Tariff and Subsidy Reform

An effective tariff and subsidy regime has to be in place in order to balance between the legitimate need to reduce the financial burdens on poor rural populations and the conditions necessary for alternative distributed electrification models to contribute to rural electrification and promote rural development. These subsidy programs have to be transparent and minimize administrative costs in order to avoid gaming of the system and to maximize benefits flowing to the intended recipients (UNEP 2002).

The subsidies themselves should have the following characteristics:

- **Clear Mandates:** Subsidies should have clear mandates and be financed by the appropriate mechanisms that are internal to the sector being subsidized. They should be based on a solid understanding of the costs and benefits of the subsidy. This is not the case in India, for example, where agricultural pumpsets are largely unmetered or the electricity tariff is not paid, effectively shifting an agricultural subsidy onto the electricity sector. This only makes it more difficult to provide electricity more widely and to effectively target and implement the subsidy (UNEP 2002).
- **Targeted:** Subsidies should be designed in a way to reach those most in need. Poorly targeted subsidies are wasted resources and do not reach the intended recipients. Subsidized lifeline tariffs based on consumption (as in Brazil) do not account for the actual financial situation within a household or even the characteristics of the household (e.g. number of people, etc.). Lifeline rates that are set well above the average level of consumption will primarily benefit those higher up on the income scale, as happened in Chad (Barnes and Halpern 2000; UNEP 2002). They can also create large financial burdens (Barnes and Floor 1996). Tying subsidies to other, well targeted, aid programs is one way to reduce administrative costs. The data from other aid programs (e.g. health, social welfare, etc.) can also be used to provide a better picture of the potential customer base in areas that are currently unelectrified.

- **Focus on First-Costs:** As a general rule, subsidies should be targeted towards solving the first-cost capital financing problem and not for ongoing consumption. The goal is to aid households and communities to transition to more modern energy systems. Subsidies for technology purchases or connections would be examples of appropriate subsidies (Barnes and Halpern 2000).
- **Phased:** Subsidies should be established with clear guidelines regarding phasing out of the subsidies. In some cases, a sunset clause based on time may the most appropriate. In other cases, subsidies could be tied to financial performance, technology development and deployment milestones or other measures of success. In all cases, however, completely open-ended subsidies should be avoided, if possible, and the rules regarding the end of the subsidy made clear in order to avoid shocks in the system once the subsidy ends. Once subsidies are in place, they are difficult to terminate and the costs can balloon, as they have in India, creating severe financial crises for service providers and governments (Barnes and Halpern 2000; UNEP 2002; Radulovic 2005).
- **Market Enhancing:** Subsidies that help develop and nurture a market early in its development can be very effective. Targeted technology assistance is one way of accomplishing this and has been used very effectively in China (Barnett 1990; UNEP 2002). If designed correctly, subsidies can lead to more competition and an improved product at lower cost for the end consumer. At the same time, subsidies can be combined with quality control programs to ensure that low quality products don't end up destroying the market in its infancy (Radulovic 2005; REDP 2006).
- **Flexible:** Subsidy programs should also incorporate flexibility to deal with the uncertainties that are an inevitable part of making changes to institutions and markets. As noted in Radulovic, and references therein, it is not always as simple as "getting the institutions right" given the various interests involved. Politics and other non-economic forces can play a large role and any subsidy program is going to have to take on those challenges as they arise (Radulovic 2005).

Changing the economics of distributed electrification requires addressing more than just subsidies within the electricity sector. Both the subsidies to other energy sources and the impact of import restrictions can affect the viability of DG options. Subsidies for electricity substitutes (e.g., kerosene, LPG) change their relative economics and make it difficult for electricity to compete (Cabraal et al. 1998; Martinot et al. 2002). High import tariffs on technology components (e.g., solar cells or diesel generators) can significantly drive up the price of distributed systems (Karekezi 1994; Cabraal et al. 1998). Taxes can also change the relative pricing of traditional versus modern fuels, creating price pressures even on those that do not get access (Barnes and Floor 1996). Tax collection requirements on electricity sales can be onerous for small entrepreneurs and raise the transaction costs of doing business.[2]

---

[2] Though, in principle, this may result not in a lack of electrification activity, but instead in activity outside of the legal regulatory regime.

Development of New Institutional Structures

Reforming the way in which distributed electrification is financed and sustainably operated has potentially serious consequences for the social contract discussed in the prior chapter. The implication of eliminating consumption subsidies entirely is that rural residents may be served with electricity that is expensive and/or of low quality and only for certain hours of the day. Some may not be able to pay those prices and be excluded from service. However, this would not address the type of equity concerns that lead to Brazil's current subsidy program and it could be politically quite difficult to remove such subsidies. Some form of lifeline subsidy would be needed at minimum. There are conditions under which cross-subsidies could be implemented while minimizing the economic damage. Such cross-subsidies, if kept to a modest level, can be effective. However, they are generally conceived of in terms of large centralized institutions that have both poor and rich consumers. If such subsidies are to be part of this vision, new transfer mechanisms would be needed that account for the political need for subsidies, the economic rationale for subsidies for those in need, and that do not preclude certain models from participating.

Transfers could occur through one of two channels. The first option would be to provide subsidies directly to the end-users. In this case, subsidies could be provided for energy in general, as has been suggested in South Africa. The advantage of an energy subsidy rather than an electricity subsidy is that it allows the consumer to make decisions based upon their energy needs and the availability of different options for meeting those needs. A more detailed examination of this option is necessary and would look at options to tie administration of this program (and qualification tests) to other social welfare programs already being implemented in rural areas. This is already done in Brazil for those consuming between 80 and 200 kWh per month.[3] Highly context specific analysis would be needed in order to design a system. One advantage of direct subsidies is that it would remove what is essentially a societal and political decision from affecting the functioning of the electric power system. This would free actors within the electricity system to make business decisions based on recovering their costs from end-users and compete on price as well as other factors. At the same time, those at the lowest end of the income scale would not lose the assistance they need to afford basic electrification.

The second option would be to create transfers among the electricity service providers either directly or via the government. This would depend on the particular institutional arrangements in each country. It could include partnerships between small actors such as NGOs, cooperatives and small entrepreneurs and the utilities

---

[3]Those consuming less than 80 kWh per month automatically get the reduced rates. Those between 80 and 200 can get a reduced rate if they are on the rolls of the social assistance programs that deliver other services.

within a regulated concession model. The possibility for such arrangements does exist within the Brazilian regulatory system. A concessionaire (the utility) can allow a permissionaire (e.g. a cooperative) to operate within its territories. However, the regulatory burdens for doing so are quite high, and it does not solve the cost-recovery problem, only the problem of the exclusive service territory. It may also be seen as an admission of failure on the part of the utility to meeting its obligations, which could make it subject to fines and penalties. This has made it an unpopular option. A mechanism to allow smaller actors to access cross-subsidization funds that come from consumers of urban utilities would also be necessary. Such transfer payment systems have not been attempted to my knowledge and a host of questions would have to be answered regarding how it would be arranged and protected from capture.

Both alternatives would also imply that all players, down to the smallest ones, would be regulated entities raising problems of the transaction costs for both parties. Regulators in the electricity sector are not used to having to deal with a large number of small regulated entities. Similarly, license applications and other regulatory transactions would have be kept minimized and simple in order to enable these small actors to be able to participate in the system. However, if such a simplified system could be put in place, it could benefit all involved. The large utilities could concentrate their manpower and resources on serving the more densely populated areas through the grid system. Smaller actors could participate fully without worrying about operating illegally or being undercut by the large utilities. Communities and individuals could opt for electrification options that are best suited to their needs without losing the lifeline subsidies. Finally, governments could meet their desired electrification targets while also supporting broader rural development goals. However, such a scheme has not been attempted and a number of pilot projects would have to be implemented. Unlike many prior pilot projects, which were designed to test suitability of a technology, these pilots would test the suitability of new institutional arrangements.[4]

In restructuring the institutional relationships between the actors in the electricity system, there will inevitably be winners and losers. Losers will not give up their privileges easily and programs have to be designed in order to align incentives and avoid both capture and obstruction (Radulovic 2005). Politics is not something that can simply be glossed over and ignored or treated as some sort of black box (Heller et al. 2003).

---

[4] Some form of backstop guarantee may be necessary to ensure that communities would continue to receive electricity even if the new institutional arrangement failed. That way areas in which pilot programs were initiated would not be penalized for their willingness to experiment with the new institutional arrangements. In places where strong universal service commitments have been made, the program may also have to be implemented in a way that maintains the obligations for universal service that the population has come to expect (e.g. in Brazil). Utilities would not be let completely off the hook if other actors were not ready to move in and take their place.

Conclusion

In the previous chapter a new version of the social contract was briefly laid out in which donors and, more importantly, governments were no longer solely responsible for ensuring cheap, universal service for all of the population. However, the challenge remains to attract the significant investment required for electrification (both centralized and decentralized) given the difficulties posed by serving these markets. The technical costs are high, the ability to pay is often low and financial resources are difficult to access. The conventional solution is based on subsidies that have the unfortunate side-effects of being inefficient, often destroying nascent markets that may be more sustainable over the long-term and creating budgetary shortfalls that governments must cover through taxes or deficits.

There are alternatives to relying solely on subsidies. If one takes a global view then a significant amount of experimentation has occurred (e.g. micro-credit, private finance, tariff reforms, etc.) but it is still too limited to have had a major effect. The trend, however, is increasingly towards more market based solutions as subsidies (and the prior alternative of simply not providing electricity to certain populations) become less and less tenable. There is a need for increased experimentation (with appropriate safeguards to minimize impacts on the most vulnerable) that involves the various actors in innovative institutional arrangements.

Combining the discussion in the previous chapter on restructuring the social contract with the discussion in this chapter on new approaches to financing, we can see that the roles of the private sector and of governments must be significantly different for sustained implementation of distributed generation to play its necessary role in solving the rural electrification problem. The government (as well as major donors) would play much more of an enabling role that allows the private sector to be creative in its solutions to the rural electricity problem and to match local solutions to local needs. By removing itself from more direct action and by reforming the tariff and subsidy structures the government can now focus on ensuring that the incentives exist for others to serve what would ordinarily be seen as difficult-to-serve customers. Governments can also now concentrate on the subset of the population that requires the most assistance and, through the reformed subsidy programs and innovative institutional arrangements discussed above, create more efficient mechanisms to ensure equity and universal service that are more likely to be successful.

References

Banerjee R (2006) Comparison of options for distributed generation in India. Energy Policy 34:101–111

Barnes DF, Floor WM (1996) Rural energy in developing countries: a challenge for economic development. Ann Rev Energy Environ 21:497–530

Barnes DF, Halpern J (2000) Subsidies and sustainable rural energy services: can we create incentives without distorting markets? Joint UNDP/World Bank Energy Sector Management Assistance Programme (ESMAP), Washington, DC, p 13

Barnett A (1990) The diffusion of energy technology in the rural areas of developing countries: a synthesis of recent experience. World Dev 18(4):539–553

Biswas WK, Diesendorf M et al (2004) Can photovoltaic technologies help attain sustainable rural development in Bangladesh? Energy Policy 32:1199–1207

Cabraal A, Davies-Cosgrave M et al (1998) Accelerating sustainable photovoltaic market development. Prog Photovoltaics Res Appl 6:297–306

de Aghion BA, Morduch J (2005) The economics of microfinance. The MIT Press, Cambridge, MA

Dunn S (2002) Micropower: new variable in the energy-environment-security equation. Bull Sci Tech Soc 22(2):72–86

Elias RJ, Victor DG (2005) Energy transitions in developing countries: a review of concepts and literature. PESD Working Papers. #40. Program on Energy and Sustainable Development, Stanford University, Stanford, CA, p 38

ESMAP (2000) Brazil rural electrification with renewable energy systems in the Northeast: a preinvestment study. 232/00. Joint UNDP/World Bank Energy Sector Management Assistance Programme (ESMAP), Washington, DC, p 114

ESMAP (2001) Best practice manual: promoting decentralized electrification investment. ERM248. Energy Sector Management Assistance Programme

ESMAP (2002) Energy strategies for rural India: evidence from six states. 258/02. Joint UNDP/World Bank Energy Sector Management Assistance Programme (ESMAP), Washington, DC, p 200

Heller TC, Tijohn HI et al (2003) Electricity restructuring and the social contract. Working Paper #15. Program on Energy and Sustainable Development, Stanford, CA

Illskog E, Kjellstrom B et al (2005) Electrification co-operatives bring new light to rural Tanzania. Energy Policy 33:1299–1307

Karekezi S (1994) Disseminating renewable energy technologies in Sub-Saharan Africa. Ann Rev Energy Environ 19:387–421

Martinot E, Chaurey A et al (2002) Renewable energy markets in developing countries. Ann Rev Energy Environ 27:309–348

Nicholson W (1998) Microeconomic theory: basic principles and extensions. The Dryden Press, Fort Worth, TX

Radulovic V (2005) Are new institutional economics enough? promoting photovoltaics in India's agricultural sector. Energy Policy 33:1883–1899

REDP (2006) NDRC/GEF/The World Bank: China renewable energy development project. NDRC/GEF/The World Bank China Renewable Energy Development Project, Beijing, p 35

REN21 Renewable Energy Policy Network (2005) Renewables 2005 global status report. Worldwatch Institute, Washington, DC, p 37

Srivastava L, Rehman IH (2006) Energy for sustainable development in India: linkages and strategic direction. Energy Policy 34:643–654

Tongia R (2007) The political economy of Indian power sector reforms in Victor D and Heller T eds. The Political Economy of Power Sector Reform: The Experiences of Five Major Developing Countries. (Cambridge: Cambridge University Press): pp 109–174

Uddin SN, Taplin R et al (2006) Advancement of renewables in Bangladesh and Thailand: policy intervention and institutional settings. Nat Resour Forum 30(3):177–187

UNEP (2002) Reforming energy subsidies: an explanatory summary of the issues and challenges in removing or modifying subsidies on energy that undermine the pursuit of sustainable development. United Nations Environment Programme, United Nations, New York, pp 1–31

van Campen B, Guidi D et al (2000) Solar photovoltaics for sustainable agriculture and rural development. #2. Food and Agriculture Organization, United Nations, Rome, pp 1–76

World Bank (1996) Rural energy and development: improving energy supplies for 2 billion people. 15912 GLB. World Bank, Industry and Energy Department, Washington, DC

Index